MODULAR SCIENCE for GCSE

MODULE 4 *Energy*

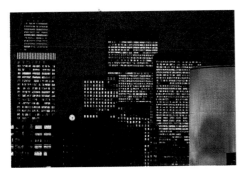

*In any process or change, **energy** is transferred. Energy is being changed from one form to another all around you in the natural and technological world. This module looks at energy in all its forms, where we get energy from and how we use it today.*

Relevant National Curriculum Attainment Targets: 13, (5)

4.1	What is energy?
4.2	Active energy, stored energy
4.3	Electricity – simplicity!
4.4	Supply and demand
4.5	Pumped storage systems
4.6	Heat and temperature
4.7	Heat energy on the move
4.8	Stopping heat from moving
4.9	Energy in the home
4.10	Fuels: from boom . . . to bust?
4.11	Coal
4.12	Oil and gas
4.13	Other sources of energy
4.14	Nuclear energy
4.15	The nuclear debate
	Index/Acknowledgements

MODULE 4

4.1 What is energy?

The reason for change

Energy is needed to make anything happen – it makes the sun shine, it makes cars and buses move, it even keeps you alive!

For something that is so important in our lives, energy is very hard to explain. We know we have to 'save it' if we want to keep our bills down. We know that the glucose in a Mars bar is 'for energy'. If somebody is very active, we might say that they are 'energetic'. But exactly what is energy? Where does it come from? What kinds of energy are there? Whatever it is, it certainly comes in a lot of different disguises!

Energy helps you work, rest and play!

1. Make a list of all the different kinds of energy shown in the three scenes on these pages.

2. For each of these kinds of energy, try to list the ways in which we can use them.

3. Write a sentence or two to explain what you now think energy is!

If you looked up the word 'energy' in a science dictionary, you would find something like 'energy is the capacity to do work'. But what does work mean? Work in this sense can mean making anything happen.

Energy is often stored in energy sources, such as fuels and batteries. But in many cases the energy from these sources is changed into another form of energy before it is used. Whatever form it's in, energy is useful stuff. You can find out more about energy by reading through this module.

4.2 Active energy, stored energy

Lots of energy

Though there are many different types of energy, it is possible to sort energy into two main forms. Some types are **active** and so are more obvious. Light can be seen, sound can be heard, heat can be felt. A moving object clearly has energy, too.

But where is the energy in a piece of coal, a sausage, a wound spring, a battery or a rock on a hillside? The energy here is **stored**, and only appears when something happens to make it *change* into active energy.

From stored to active

Some stored energy is called *potential* energy because it has the '*potential* to do work'. For example, things that are high up have **potential** energy. If they are not held in place they will fall downwards because of the pull of gravity. As they fall, they lose this potential energy, which is turned instead into energy of motion, called **kinetic** energy.

Other energy 'stores'

Energy can also be stored in things that have been bent, squeezed or stretched.

A bent bow has stored energy – when it is released, this turns into the kinetic energy of the moving arrow. This is similar to the energy stored in a stretched spring or elastic.

The wound spring in a clockwork motor stores energy in much the same way. A spring metal strip is coiled tightly by winding a key, or by turning the wheels. When released, the metal tries to straighten out. This releases the stored energy, causing the wheels to turn – releasing kinetic energy.

Because these ways of storing and releasing energy all involve *mechanisms*, this type of stored energy is called stored **mechanical** energy.

Stored chemical energy

Some materials have energy locked up in the chemicals from which they are made. Perhaps the simplest example of this is coal. A lump of coal may not seem very energetic, but just try burning it to make it react with the oxygen in the air! A lot of energy is given out in the form of **heat** and **light** energy. Materials like this – such as wood, coal, oil and gas – are called **fuels**.

The food we eat reacts in a similar way inside our bodies. The energy released keeps us warm and also provides the energy we need for life – food is like a fuel for our body!

Batteries also contain stored chemical energy, but this time the chemicals react to give **electrical** energy. This is a particularly useful type of energy as it can be easily turned into many other types of energy, such as light, sound, heat...

At a barbeque, chemical energy stored in the charcoal is released as heat energy. The cooked food also releases its stored chemical energy – inside your body.

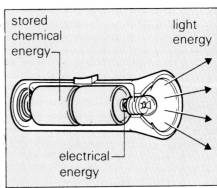

From active to stored

Look at the roller coaster shown. The potential energy it has at the top of the hill is turned to kinetic energy as it rolls downhill. But it is turned back to potential energy when it goes uphill. So you are storing potential energy whenever you lift something.

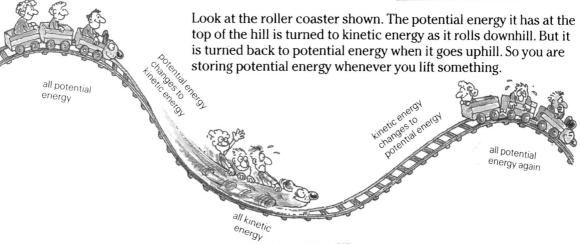

A similar process happens in a car, but involves electricity. When a lead-acid (car) battery is connected to the headlamps, the energy stored in the battery is turned into electrical energy and then into light energy. But once connected to a battery-charger, electrical energy is pushed back into the battery, refilling the energy store!

1 List some examples of active and stored energy.

2 What are **two** main types of stored energy? For each, try to explain how the energy is stored.

3 What type of energy does stored mechanical energy usually turn into?

4 What types of energy can stored chemical energy turn into?

5 How can active energy be turned back into stored energy?

4.3 Electricity – simplicity!

Non-stop energy

As we have seen, the different types of energy can be changed from one to another. In fact, when we say we 'use' energy, it is not 'used up', it is simply changed from one form to another. Energy cannot be created or destroyed – it can only be released or stored.

Some changes are easier to achieve than others and special devices are often needed to make the change. Electrical energy is generally the easiest form to change, which is why we use so much of it!

Energy can never be lost – it just changes into a different form. But it is easily wasted!

The electrical merry-go-round

For every device which changes electrical energy into another form of energy, there is usually another device which can convert that energy back into electrical energy.

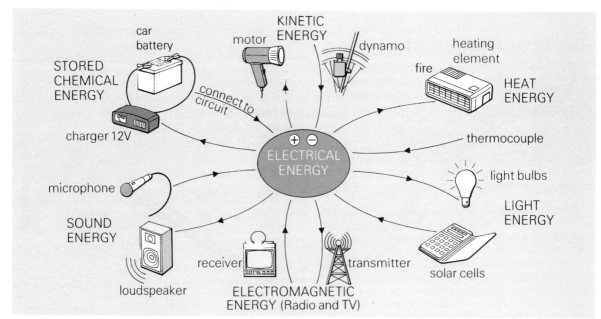

Measuring energy – the joule

Unlike distance or time, energy is not usually easy to measure – there is no simple 'energy' equivalent of a ruler or stopclock! The unit of energy is the **joule** (J). One joule is about the amount of energy needed to lift a small apple from floor to table top. However it is possible to work out the energy used by electrical equipment fairly easily.

If you went out to buy a light bulb, you could choose from various bulbs, such as a 60 or 100 watt bulb. The number of watts tell you how bright the bulb will be. These figures are a measure of the **power** of the bulb – this is the rate at which things use energy. For every **watt** of power, *1 joule of energy is being used every second*. So a 100 W bulb uses 100 joules of energy per second.

A 1000 W fire uses as much energy each second as ten 100 W light bulbs.

Power to the people!

In an industrial country like Britain, there is a large demand for electricity. In all, there are 81 **generating plants** (power stations) in Britain. That's enough to keep over 500 million light bulbs constantly lit.

Although these generating plants use different energy sources, most of them use a long 'energy chain' like this.

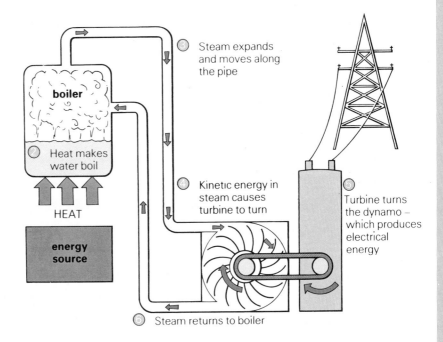

① The stored energy in the energy source is released as heat energy
② Heat makes water boil
③ Steam expands and moves along the pipe
④ Kinetic energy in steam causes turbine to turn
⑤ Turbine turns the dynamo – which produces electrical energy
⑥ Steam returns to boiler

But is it efficient?

The trouble is that every energy change is less than perfect. Not all the energy released by the energy source ends up as electricity. A lot ends up as 'waste heat' which never reaches the turbine. For every 100 joules of 'stored energy' that start the journey, 70 joules are wasted and only 30 get as far as your home! Because of this, electricity, though very useful, is an expensive way for us to get our energy!

Household waste?

Once the electrical energy gets to our home, do we waste it too? The answer is yes and no. An electric fire can be very efficient – producing 95 joules of heat from every 100 joules of electrical energy. But the humble electric light bulb is an efficiency disaster – only 5 joules of light energy from every 100 joules of electrical energy – the rest is heat! That's why you may need gloves if you have to change a bulb after it's been on for a while! Fluorescent tubes are much more efficient – 20 joules of light energy are released by every 100 joules of electrical energy. That's why your school uses them.

A light bulb produces a lot of heat – which is no use to see by and it may hurt too!

1. Why is electricity such a useful form of energy?

2. What is the unit of energy? How much energy is used by a 60 W bulb every second?

3. Describe and explain the basic 'energy chain' used in a power station.

4. Why does only such a small amount of the original 'stored energy' end up as electricity?

5. a Why do schools tend to use flourescent tubes rather than light bulbs?
   b Write down three reasons why people prefer to use light bulbs at home?

4.4 Supply and demand

Keep going, don't stop

Electrical energy is very useful, but it is expensive to produce. One reason for this is that oil- or coal-burning plants have to be kept running 24 hours a day! If shut down for anytime, the boilers have to be cooled slowly and carefully to avoid damage. A period of maintenance is then needed, followed by an equally slow warm-up period when they are put back into service.

The National Grid

Over a long term, the slow cooling and start up of plants is not too much of a problem. This is because Britain has an energy 'highway' system – the **National Grid**. For example, as overall demand falls in the summer, a plant at Southampton can be shut down for a few months. In the meantime that region can 'feed' off the rest of the National Grid.

'I want it now!'

The main difficulty in organising a national distribution of electricity is the ever-changing demand for electricity. At night, most of the country stops work, goes to bed – and demand for electricity drops like a stone!

But the generating plants can't just be switched off, so what can be done with all the 'surplus' electricity at night? There are two main ways of tackling the problem – either use more electricity at night or build power stations that respond more quickly.

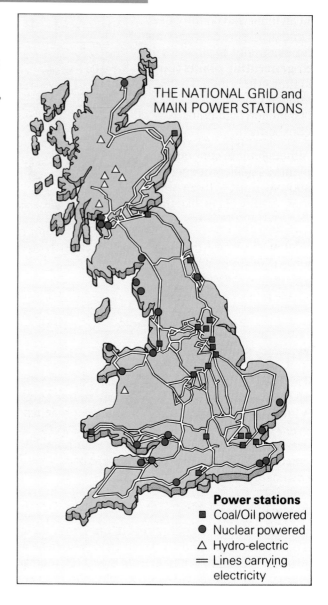

THE NATIONAL GRID and MAIN POWER STATIONS

Power stations
■ Coal/Oil powered
● Nuclear powered
△ Hydro-electric
= Lines carrying electricity

Buy now, use later!

As many people use electricity for home heating, it's a good idea to try to store the electricity as heat. **Night storage heaters** are connected to a separate payment meter that only works at night. Because electricity is plentiful at night, it is sold off at 'bargain' rates! Night storage heaters use this cheap electricity to heat up special bricks inside the heater. This heat stays stored in the bricks all through the night. This stored energy is then slowly released into the house the following day. But despite plenty of publicity, this has not really made much of a dent in the problem. This is because the main demand for electricity is when people are awake – working, cooking and watching TV. There is no way most people will do these things all night instead!

Hydroelectric plants

These do not use an energy source which produces heat – so they do not have long cool-down and warm-up periods.

Water high in the mountains has lots of potential energy which turns to kinetic energy as it flows downhill. This can be harnessed to turn turbines, spin dynamos and so produce electricity.

In countries with high mountains and lots of rain, like Scotland and Wales, high valleys are often dammed and their water is chanelled through pipes to turbines to generate electricity.

A hydroelectric plant can be shut down or started up in minutes rather than days so they are very useful plants to include in the National Grid. Hydroelectric plants in mountainous areas of Britain can be used to supply electricity all over the country. Another advantage to these plants is that the rain comes free! So not only are they flexible but they are cheap to run too!

The natural conditions here in the north of Scotland – high valleys, plenty of rain – are ideally suited to the building of hydroelectric plants like this one. Other areas also have natural conditions which can be used to generate a lot of electricity in a similar way. Tidal estuaries, windy plains and hillsides are examples of such safe sources of power.

1. How are long term changes in the demand for electricity coped with?

2. Why are the short term changes (day/night) such a problem?

3. What are the energy changes that take place in a hydroelectric plant? Draw a simple energy chain diagram to show all the changes.

4. 'What use are night storage heaters? They heat up at night, when I'm already warm in bed'.
How would you convince this person that such heaters are useful?

5. Explain why a hydroelectric plant is a useful type of power station to include in the National Grid.

4.5 Pumped storage systems

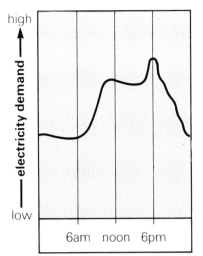

At what times will sudden increases in demand mean that the National Grid will need electricity from Cruachan?

A combined effort

Recently, various projects have brought together hydroelectric plants and a new method of storing electrical energy. Take a closer look at one of these projects...

The Cruachan Dam holds back a vast lake high in the hills of Argyll, Scotland. This is the potential energy storehouse. It is added to by rainfall over the hills (which is common!) but it also has another way of filling.

During the peak daytime demand in electricity, the plant acts as in a 'traditional' hydroelectric plant. Water falls through tubes the size of railway tunnels to the generating hall – a great cavern cut from solid granite deep in the mountain. There it is used to drive the special turbine-dynamos which then feed electricity into the National Grid – all within minutes of starting the plant up.

But the water does not flow away – instead it is kept in a lower reservoir. At night, the National Grid is full of surplus electricity, but few people are awake to use it. Now the machinery goes into reverse. The turbine-dynamos become motor-pumps which push water back up into the upper reservoir – storing some of the unwanted electrical energy as potential energy. This water can then be used to generate electricity in the normal way – such as during the next day when everyone is awake and at work.

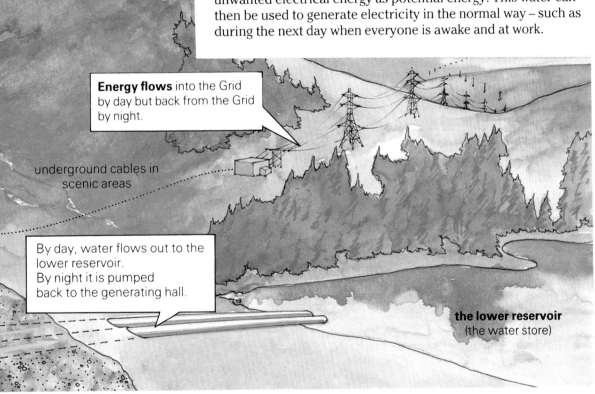

Energy flows into the Grid by day but back from the Grid by night.

underground cables in scenic areas

By day, water flows out to the lower reservoir.
By night it is pumped back to the generating hall.

the lower reservoir
(the water store)

1 List all the energy changes involved in turning the potential of the water in the dam into electrical energy.

2 You are an engineer who wants to build a pumped storage plant in the mountains, but local people fear that you would damage the environment. Set out your case for building the plant. Explain briefly how it will work and why it is needed. What steps will you take to protect the environment?

4.6 Heat and temperature

Getting into hot water!

'Don't touch it, it's hot!' – a common enough warning at dinner time. Everyone knows what it means. But people often get confused over the difference between heat energy and temperature.

Heating things up

If we put heat energy into something (by burning a fuel or using electricity), it gets hot – its temperature rises. **Temperature** is a measure of how hot something is. But a given amount of heat energy does not always produce the same rise in temperature if different materials are heated.

Some materials need to absorb more heat energy than others to give the same change in temperature. The amount of material being heated also makes a difference.

A full kettle will take twice as long to boil as a half-full kettle.

Heat, temperature and moving energy

If we talk about the amount of **heat** energy that an object has, we are actually describing the amount of **'internal'** energy that an object has. This depends on what the material is, how much of it there is, and how hot it is. One way to think of *temperature* (hotness) of an object is as a description of how *concentrated* that heat energy is in the object.

In a bucket of cold water, the heat energy will not be very concentrated – yet the total amount of energy in the whole of the bucket will be quite large. A sparkler 'spark' is very hot and so its heat energy is very concentrated – yet there is only a tiny amount of material in the spark – so the total amount of energy is small.

So which way will energy flow if a spark lands in the water? In fact, the spark 'loses' energy and cools down (a lot) while the water gains energy and warms up (a little!). This cooling and warming carries on until they are both at the same temperature. Heat energy *always* flows from hot things to cooler things until they are both the same temperature.

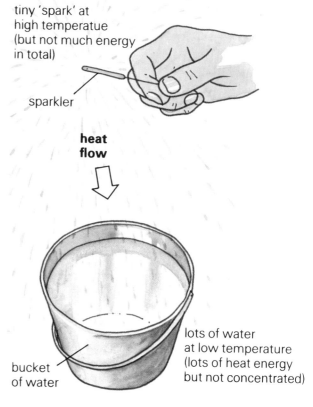

What will happen when a spark hits the water?

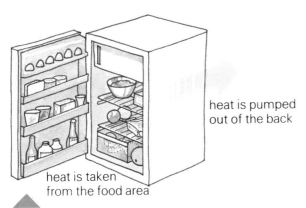

heat is pumped out of the back

heat is taken from the food area

▲ Fridges feel warm at the back as heat energy taken from the food is released.

How do things get cold?

The simple answer is by passing on their heat energy to other objects. If there is some way of *continuing* to pass on heat energy, then a cold object can become even colder.

A fridge does this by circulating a cooling fluid inside the fridge. This fluid keeps coming back to remove heat energy from the food – so the food becomes cold.

Measuring temperature

Though we can feel if things are hot or cold, this is not very accurate. Our senses are easily confused – a cool swimming bath feels warm if you've just had a cold shower! We use **thermometers** to measure temperature accurately.

A typical thermometer is made from a thin-bore glass tube with a reservoir bulb at one end, full of mercury (or coloured alcohol). If the bulb is heated, the mercury **expands** (gets bigger). The glass also expands, but not as much as the liquid. The expanded liquid pushes up into the tube. The hotter it gets, the further it goes – if it cools, it **contracts** (shrinks) back down the tube.

The tube is marked with a scale by marking the points for melting ice and boiling water (taken as 0 degrees and 100 degrees), and dividing the rest of the tube evenly between these **fixed points**. This is called the **Celsius** scale, after its inventor. Each point on the scale differs by one degree **centigrade** (°C).

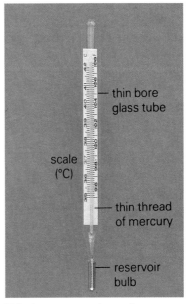

This thermometer is used to measure body temperature. If this is too high or too low, a person is ill.

Why we get cold

Our normal body temperature is 37°C – that's hotter than the hottest summer day! A warm room is at about 20°C, so even then our bodies constantly lose heat to 'warm the room'. We replace this heat loss with heat from our fuel – food! The more the temperature drops, the more we have to use up our food to keep our body at 37°C. Sometimes our muscles do a little extra work to help this – by shivering. But if it gets too cold, we may not be able to generate heat fast enough and our temperature will fall below 37°C. This is called **hypothermia** and can be fatal – especially for the old or very young.

1 Which would boil first over the same flame – a full kettle or one that is only half full?

2 a Which has more heat energy, a sparkler 'spark' or a bucket of water?
b What happens if they meet?

3 Describe how a simple thermometer is made. How is a scale added?

4 a Why do we shiver when cold? What good does it do?
b Suggest different ways that we can keep our temperature at the correct level.

4.7 Heat energy on the move

How heat energy moves

Given enough time, heat energy will flow from a hot place to a cool place until the temperatures have levelled out. How long it takes for this 'levelling' to occur will depend on the **temperature difference**, and also on the **materials** involved. Some let the energy flow quickly; others slow it down. To understand how best to use them, we first need to know about the three ways in which heat energy can move: **conduction, convection** and **radiation**.

Heat moves in three different ways – which one will make these balloons rise?

Conduction

When you first prod a fire with a metal poker, one end could get 'red hot' while the other end stays cool enough to hold. But after a while, the heat energy moves steadily along the poker, each bit of material being heated in turn by its neighbour. This is the way heat passes through *solids* – it is called **conduction**.

Metals are generally good **conductors** – and some are very good. A copper 'poker' would give you burnt fingers in minutes. Not all solids are good conductors. Plastic and wood are used as saucepan handles because they do *not* conduct heat well. Poor conductors are called **insulators**.

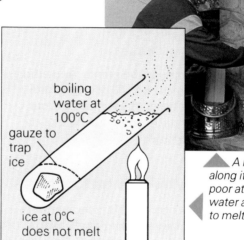

A metal soon conducts heat all along its length – but liquids are very poor at conducting heat. The hot water at the top cannot transfer heat to melt the ice.

Convection

Although liquids are poor conductors, they will let heat travel in a different way. This happens when a liquid (or a gas) is heated from below. First the liquid at the bottom is heated directly by *conduction*. This *hot* liquid **expands** and 'floats' up through the cooler liquid.

When the hot liquid reaches the top, far away from the heat source, it cools. More hot liquid (rising from below) pushes the cooling liquid out of the way. The *cooling* liquid now **contracts** and starts to 'sink'. Back at the bottom, it is heated once more and starts to rise again...

This process is called **convection** – and this flow of liquid is called a **convection current**.

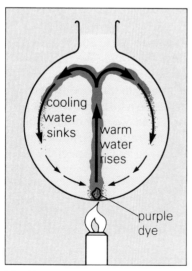

Liquids are very good at carrying heat upwards – but by convection, not conduction.

Radiation

Conduction and convection both need a material which can transfer heat energy – solids for conduction, liquids or gases for convection. So how come we feel heat from the sun, which reaches us through 150 million km of the empty vacuum of space? The answer is that the energy reaches us as **infra red radiation**, which is like light and radio waves. You can feel these infra-red waves when you warm your hands in front of an electric fire. **Microwaves** are very similar, but they have the neat trick of penetrating your dinner and heating inside and out at the same time.

By directly heating the inside of food, microwaves cook food very quickly.

Home heating on the move

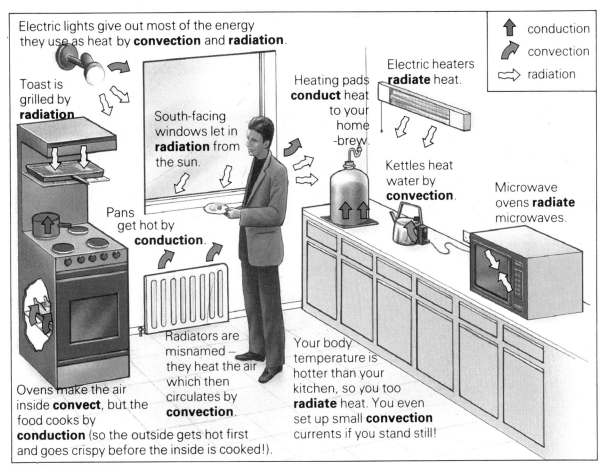

1. What are the three ways in which heat can move?

2. **a** What materials (if any) are needed for each heat transfer?
   b Describe how the heat energy moves in each case.

3. Why can't heat from the sun reach us by conduction or convection?

4. List examples of conduction, convection and radiation that you use in your home.

4.8 Stopping heat from moving

Don't waste it!

Heat energy is only useful if it can be kept where it is wanted. If heat energy moves away from such places, it is wasted – and this can be expensive!

To insulate properly we have to stop all three methods of heat transfer. The classic example of this is the vacuum flask.

plastic/cork – good insulator, doesn't conduct heat away; stops convection of hot air above liquid; blocks radiation.

double-walled – allows vacuum to be made.

hard case – protects fragile glass walls.

vacuum – no solid, liquid or gas, so there is no conduction and convection.

silver surfaces – reflects radiation back into flask.

seal point – where air was removed when making vacuum.

hot drinks stay hot (or cold drinks stay cold)

A vacuum flask can keep heat in – or keep it out – by stopping conduction, convection and radiation.

Stopping radiation

Silvery surfaces do not absorb heat radiation – they reflect it, just like a mirror reflects light. They are **poor absorbers** of heat. The larger the area of the surface, the more points there are from which it can radiate heat. Silvery surfaces are very smooth and flat. Rough surfaces have small scratches – these expose more of the surface. This means rough surfaces can radiate a lot of heat, but silver surfaces do not. So silvery surfaces are **poor radiators** of heat.

That's why your Chinese take-away stays hot in its shiny aluminium container – the inside surface doesn't absorb much heat from the food, and the outside surface doesn't radiate much heat out.

'Space blankets' are thin silvery plastic sheets. These can help to keep in your body-heat if you are out in cold weather. They can also help old people beat hypothermia (see 4.6).

Another way of *stopping* radiation is to *avoid* using **black surfaces** – these are very **good radiators**. They are also **good absorbers**, that's why a black coat feels hot on sunny days.

Marathon runners can get very cold when they stop running. 'Space blankets' help to stop them radiating heat.

Stopping convection

Like any gas, air is a very poor conductor. If it can be stopped from moving around (and so convecting), air can make a good and cheap insulator. That is how clothes and fur help to keep us and animals warm. Even the tiny hairs on our bodies stand up when we're cold, to try to trap more air and so keep us more insulated. A similar effect is given by air trapped between the sheets of paper used to wrap our fish and chips.

Other natural insulators, such as cork, trap air in tiny pockets or bubbles. Expanded polystyrene, fibre glass, string vests and foam rubber all insulate in this way – so does corrugated cardboard, as used by all the best tramps!

The fur of the arctic fox slows down convection – and its white colour reduces heat loss by radiation too!

Insulating materials for the home

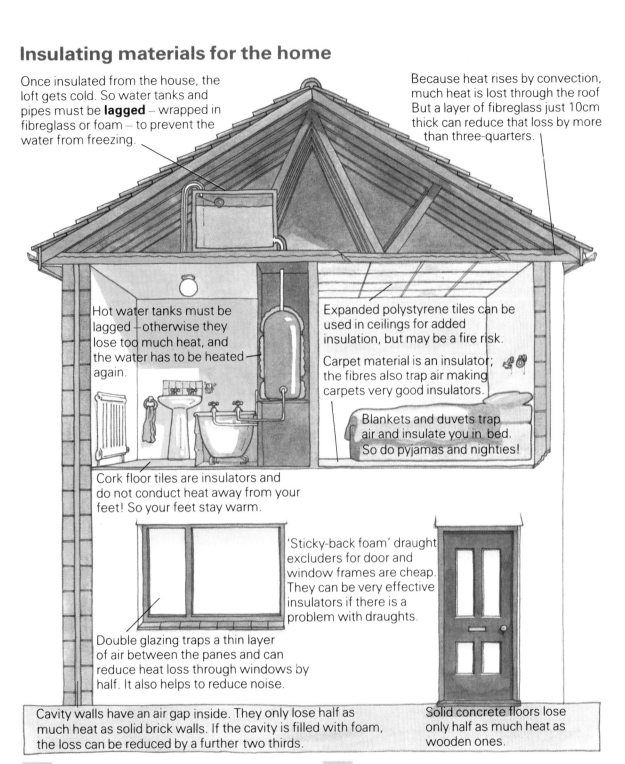

Once insulated from the house, the loft gets cold. So water tanks and pipes must be **lagged** – wrapped in fibreglass or foam – to prevent the water from freezing.

Because heat rises by convection, much heat is lost through the roof But a layer of fibreglass just 10cm thick can reduce that loss by more than three-quarters.

Hot water tanks must be lagged – otherwise they lose too much heat, and the water has to be heated again.

Expanded polystyrene tiles can be used in ceilings for added insulation, but may be a fire risk.

Carpet material is an insulator; the fibres also trap air making carpets very good insulators.

Blankets and duvets trap air and insulate you in bed. So do pyjamas and nighties!

Cork floor tiles are insulators and do not conduct heat away from your feet! So your feet stay warm.

'Sticky-back foam' draught excluders for door and window frames are cheap. They can be very effective insulators if there is a problem with draughts.

Double glazing traps a thin layer of air between the panes and can reduce heat loss through windows by half. It also helps to reduce noise.

Cavity walls have an air gap inside. They only lose half as much heat as solid brick walls. If the cavity is filled with foam, the loss can be reduced by a further two thirds.

Solid concrete floors lose only half as much heat as wooden ones.

1 How does a vacuum flask stop conduction, convection and radiation?

2 Why does take-away food come in aluminium containers? Why does the food stay hot?

3 Why do tramps often wrap themselves in cardboard boxes and newspaper when "sleeping rough" in the cold?

4 Make a list of the different insulators used to keep a house warm.

5 Draw a plan of your own home. Show how heat may escape and suggest ways to stop the heat loss in each case.

4.9 Energy in the home

What's it used for?

The main uses of energy in the home are for heating, lighting, cooking and for electrical appliances. The electrical appliances you use are up to you. But in Britain, everyone needs some form of heating, lighting and cooking. And there is more than one way of doing each.

Hot house

Long ago, rooms used to be heated by an open fire in each room. But each open fire let a lot of heat escape up the chimney. So **central heating systems** were developed. These used just one source of heat – such as a boiler heated by **gas**, **oil** or **coal**. Only one chimney is needed, a special one which does not allow much heat to escape. The boiler heats water which then flows through pipes to radiators. These radiators then warm up each room. The water then goes back to the boiler to be heated again. Most boilers also heat up a separate supply of water for the hot taps.

Electric central heating is not really central heating – it uses separate night storage heaters (see 4.4). Each heater produces heat from its own supply of electrical energy – there is no central source of heat energy. Hot water for taps comes from a water tank heated by an electrical **immersion heater.**

An alternative to central heating is to use *individual* sources of heat – such as a fire (coal, gas or electric), a fan heater or a portable heater (electric or gas). The kitchen or bathroom may have a wall-mounted water heater. These sources are often more suitable for a particular room than a radiator would be. They are frequently used to give extra heat, in addition to a central heating system.

Light bills?

For a particular amount of electricity, fluorescent lights (strip lights) give out four times more light than electric light bulbs. But many people do not like to use them in their homes – and choose to use the less efficient electric bulbs. This is because lighting does not form a big part of their energy costs.

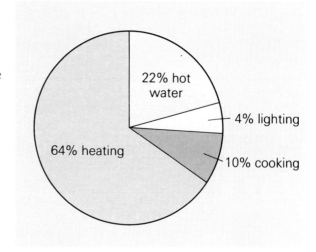

Energy used in the home for heating, cooking and lighting.

coal
(cheap)

needs regular 'top ups' and cleaning, pollution risk from mines, soot and gases

oil
(cheap)

easy to use, pollution risk from oil slicks

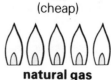
natural gas

very cheap, but not available everywhere, no pollution

LPG

'bottled gas', cheap, portable, no pollution

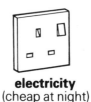
electricity
(cheap at night)

expensive by day, easy to install, pollution risk from acid rain, nuclear waste

Different fuels used in the home.

Making a meal of it

There are many different ways of cooking – using an oven (gas or electric), a grill, a pressure cooker, a slow-cooker, a toaster, a microwave oven and so on. Each method has its advantages. Microwave ovens are very efficient at cooking food in small amounts, or which contains a lot of water. If lots of food is to be cooked, then a conventional oven may be more cost-effective. You just have to choose a method that suits your needs and your purse!

Cheap, quick and effective insulation – draught excluders.

Hidden insulation – foam can be injected to fill the hollow cavity in most home walls.

Spend money to save money!

It costs money to produce the heat energy needed to warm a home. In time, all of this heat escapes – that is why a home eventually gets cold. If any of this heat escapes too quickly, then money is wasted along with the heat. So even a small reduction in the amount of heat wasted can mean big savings.

Let's consider a home with a heating bill of £300 a year*. If 10% of the heat can be kept in for longer, then the heating bill will be reduced by £30. But how much money should be spent on insulation? It all depends on which type of insulation is chosen...

Heat escapes through...	Percentage of heat escaping	Cost of heat escaping*	Possible insulation	Reduction in heat escaping	Cost of insulation	Years to recover cost	Comments
Walls	35%	£105	foam in cavity walls	35% → 14%	£315	5	needs experts
Roof	25%	£75	fibre glass in loft	25% → 5%	£180	3	cheap, easy to do
Draughts	15%	£45	draught excluders	15% → 6%	£54	2	very cheap, very easy
Windows	10%	£30	double glazing	10% → 4%	£216 or £1800	12 or 100	do-it-yourself or expert fitted

More than just money

There are other considerations besides the cost. Many people like to have open fires instead of modern forms of heating. But a central heating system can be operated by timers which make heating a home very easy and convenient – the choice is yours!

1 Why is central heating so called?

2 Which form of insulation is cheapest and easiest to put in?

3 Make a list of the advantages and disadvantages of the different types of home heating.

4 If you had £400 to spend on insulation, how would you spend it? Explain your answer.

4.10 Fuels: from boom... to bust?

The good old days

For hundreds of years, coal was a very important fuel. Not only could it be used for heating, but it was also the source of many chemicals for industry. By the 1960s, coal had been replaced by **oil**. Oil was used in power stations, for petrol, for plastics ... the list of uses was very long indeed.

Many countries rode the crest of the 'cheap energy' wave, thanks to the glut of oil from the Middle East. People who suggested we were wasting a very precious resource were laughed at ...

How cheap was cheap? Find out how much 7/6 (7½ shillings) is in pence. How much is petrol now?

The crisis

Then, in the early 1970s, the main group of **o**il **p**roducing & **e**xporting **c**ountries (OPEC) limited oil production and fixed a higher price. They realised that they only had a limited amount of oil to sell. They had to make it last and get as much money for it as possible.

In countries like Britain which imported a lot of oil, the result was economic chaos. Petrol prices rocketed. So did the cost of electricity because oil was used in many power stations. Industry used vast amounts of oil too, so the 'knock-on' effect of high oil prices pushed up the price of many other goods.

An OPEC conference – these representatives of the main oil producing countries can decide the price for oil world-wide.

Fuels last longer and bills get smaller if you 'save it'!

The need for conservation

With this rise in the price of oil suddenly it became very important to use less fuel and save money. In the '70s cars began to be designed and sold for their 'fuel economy' rather than their 'performance'. People became more aware that the **fossil fuels** (coal, oil and gas) would soon be used up and could not be replaced. In the short term there was only one thing to be done – try to 'save it' – try to economise on our comsumption of fuel to make what we had last longer.

The good news . . .

In Britain, because imported oil was now dearer, it was worth spending more money to look for it! Billions of pounds were spent searching for oil in the rocks beneath the **North Sea**. Very soon this money was pouring back into the economy when the North Sea oil rigs started to produce high quality oil. By the early 1980s, Britain had more than enough oil to use at home and became an oil exporting country!

Drilling for oil – the precious 'black gold'.

. . . or is it?

But by 1985 things began to totter. The energy-saving campaign, coupled with the general industrial slump, meant that there was a lower demand for oil worldwide.

At the end of the year, the OPEC countries failed to agree to reduce their production evenly. Once again there was an oil surplus – and prices fell drastically. The energy consumers of the world were pleased. Petrol could go down in price, electricity might be cheaper, lower fuel costs would be 'good for industry' and so would help the economy. . .

The only problem was that Britain was also an oil exporter. Falling oil prices meant getting less money for the oil we sold.

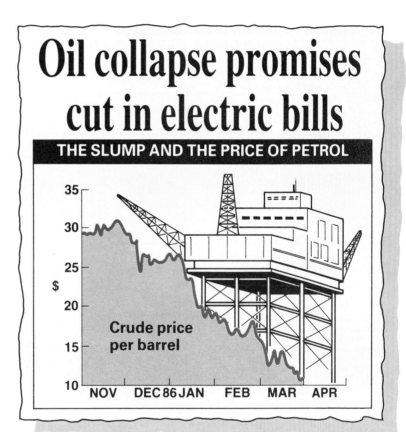

What next?

Eventually the oil will run out. The flow of North Sea oil is already slowing down. So where will we get our energy supplies from? **Nuclear** energy is one possibility but so are **alternative** energy sources. As importantly, where will we get a new supply of the chemicals we obtain from oil? One source is plants, using **biotechnology** to extract useful chemicals. Another source, which should last for 300 years is **coal**. Back to where we started!

1 Suggest five ways that we could all save energy.

2 It is easier to use the oil we have now than it is to find new oil supplies. Do you agree? Explain your answer.

4.11 Coal

Plants – natural energy stores

We use plants as 'body fuel' (**food**) but we also rely on them as sources of energy for our technological society. In the simplest form, we burn **wood** from trees to keep us warm. The trees have energy (from sunlight) stored in the complex chemicals that make up wood. By reacting these chemicals with oxygen – burning them – we can get back some of that 'sunlight energy'.

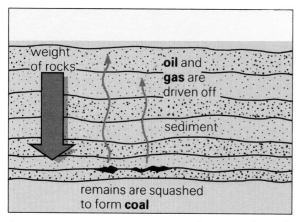

Great heat and pressures deep in the Earth transformed buried plant material into coal.

Fossil fuels – ancient energy stores

This process of 'trapping sunlight' has been going on for millions of years. Most of the plant material formed in this way broke down soon after the plants died. It then released its energy to bacteria and fungi that cause decay.

But some plant material became trapped in sediments. This material was sealed in and protected from total decay. This material has retained its trapped energy – until now, when we dig up this **'fossil'** material and use it as fuel. When we burn fossil fuels – **coal**, **oil** and **gas** – we are releasing the energy store that was built up by plants, using sunlight, millions of years ago.

Millions of years in minutes!

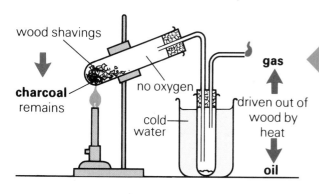

This experiment shows how fossil fuels might have formed when plants (and animals) were buried by sediments, millions of years ago.

Wood is heated in a closed container, away from the oxygen that would make it burn.

As the wood is heated, any liquids and gases in it are driven off and the complex chemicals break down. What is left behind is charcoal – almost pure carbon. Coal must have formed, deep underground, in a similar way.

British coal

From soft plants to hard rocks – powerful cutting machines are needed to remove the coal.

Britain is very lucky in having large coal reserves. This is because about 300 million years ago, it was largely covered by great swamp forests. Rivers rushing from mountains in the north poured sediment into the area, burying great areas of forest that had sunk beneath the water. This process was repeated many times. Eventually the forests were crushed and the many British coal fields were formed.

From plants to coal

It has taken 300 million years of deep burial to turn the swamp forests' trees to coal, but plant material has collected in a similar way ever since. How much this material has changed depends on how old it is.

Peat In the moorland areas of Scotland and Ireland, the remains of *thousands* of years of 'bog moss' have been compacted to form **peat**. This soft brown material still shows traces of tougher roots and leaves. It is cut with a spade and allowed to dry, and then used as a fuel (as well as being spread on gardens!). It is not a very high grade fuel, though, as it burns at a low temperature and leaves a lot of ash.

Lignite This is older plant material that has collected over the last *100 million* years. It has been compressed more than peat, and has had more of its 'volatiles' driven off. What is left is a brown, soft, waxy material called **lignite**. This can be cut into thin strips and burnt like a candle. This 'halfway' coal is found in Yugoslavia and Greece and a large deposit is now being mined beneath an Irish loch.

Bituminous coal This is typical 'household' coal. It may be hard, dark and shiny, or soft and powdery. Often it is a streaky mixture of the two types. It has a much higher carbon content than lignite, and needs to be given a 'kick' of energy to get it burning. Once it is burning, it produces a hot, but smokey flame.

Anthracite This is a very 'high-grade' coal (nearly pure carbon) that is found where coal seams have been folded or heated deep in the Earth. It is hard and shiny and doesn't mark the hands. **Anthracite** is also hard to light but, once lit, burns with a very hot and smokeless flame, leaving little ash. For this reason, it is used in solid fuel central heating systems. It can also be burnt in **smokeless zones** (such as London).

1. Where did the stored energy in fossil fuels originally come from?

2. Why has Britain got such a lot of coal?

3. Describe the differences between Peat and Anthracite.

4. If peat were buried for 300 million years, what might it end up as?

5. **a** Which type of coal must be burnt in smokeless zones?
   b Why do you think it is necessary to have smokeless zones?

4.12 Oil and gas

A long time coming

Oil and gas were produced over millions of years from dead plant and animal materials. These materials often collected at the bottom of ancient seas and were covered by layers of sand and mud. This prevented them from decaying – instead, great heat and pressure broke the materials down into **crude oil** and **natural gas**. The tiny droplets of oil and bubbles of gas rose through sponge-like **porous rock** like sandstone. In some places, they kept on rising until they reached the surface. The black oil collected there as a black sludge.

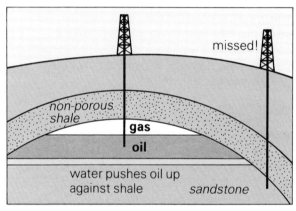

Oil was first found at the surface.

Black sludge to black gold

Nobody bothered with it until 1859 in Pennsylvania, USA where it was discovered that the black sludge could be made into a good substitute for the whale oil used in lamps. Soon people began drilling all over for the crude oil. By **purifying** the sludge, different parts could be used for heating, lighting and water-proofing. Soon it was in demand as a fuel for cars. The oil rush had begun! Nowadays, oil is also used to make plastics and other useful chemicals.

Most oil and gas is trapped under dome-shaped rocks.

Oil traps

Once all the oil near the surface had been used up, scientists had to look for it deeper in the ground. Sometimes the dead plant and animal material had been covered up by mud millions of years ago. This mud eventually formed **non-porous shale** which trapped the oil and gas. All the scientists had to do was to find the right rock formation, drill a hole and pump up the oil and gas. Sounds easy, but it costs millions to drill one well – and only 1 in 40 wells turn out to be in the right place!

The many dome shapes on the surface of this area suggest oil may be below – the many small oil wells confirm it!

What's it good for?

Each fossil fuel has different uses – but most involve releasing the energy locked up in the fuels. During burning (**combustion**), this energy is released as **heat** and **light** energy. For example, a flame is just a collection of hot gases giving off light energy. As well as heating homes, the heat energy can be used to make gases expand in engines. These gases push on pistons, in the engine, **making the machine move.**

What happens during burning?

Oil and **natural gas** are made up of chemicals called **hydrocarbons**. These contain various amounts of **hydrogen** and **carbon**. When they are burned, the chemical energy stored in the hydrocarbons is released as light energy and heat energy. It is not always easy to start the reaction – a small amount of heat is needed to get things going. Usually a spark or flame is enough. Once it is burning, the fuel itself can provide the energy needed to keep the reaction going.

One fossil fuel – **coal** – is *not* a hydrocarbon. It is almost pure **carbon**. It burns to give only carbon dioxide, plus light and heat energy. Sometimes other chemicals are present in the fuel. These produce **pollutant gases**, such as **sulphur dioxide**, which can cause acid rain. If there is not enough oxygen, the fuel does not burn properly and a poisonous gas called **carbon monoxide** is produced. This is why it is important to keep fires well-serviced and well-ventilated.

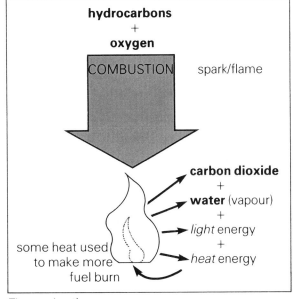

The combustion process.

What's the difference?

Solid, liquid and gas fuels burn in different ways...

Coal – most is 'hidden' inside the lump, away from air. Burns with air only on surface, so takes time to get hot enough to keep combustion going. Made from tightly packed carbon, each lump releases a lot of energy.

burning coal

Oil, petrol – if sprayed in small drops, it burns quickly. Burns with air on surface of *each and every* drop. Together, drops soon release enough heat to make fuel burn quite quickly. Liquid fuel may also form a vapour which can burn very easily. Not as tightly packed as coal, so releases less energy than small volume of coal.

burning oil droplets

Gas – mixes very quickly with air. Rapidly releases enough heat to keep gas burning. Not at all tightly packed, so large volume needed to release a lot of heat.

burning gas

 What are the three main uses of crude oil?

 Why do oil and gas collect under shale?

 Why is a spark or flame needed to start a fire?

 Water can absorb a lot of heat. Suggest a reason why water can put out a fire.

4.13 Other sources of energy

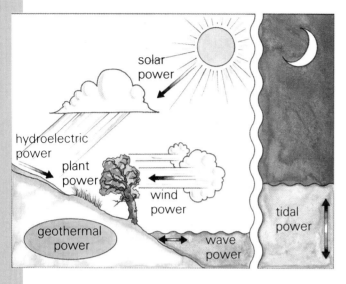

What's the alternative?

With fossil fuels running out, one day we will have to rely on other sources of energy. It may seem a long way off, but if we start using other sources of energy *now*, the remaining fossil fuels will be used less quickly. The other sources of energy may be able to fill the role of a fossil fuel as an *energy* source. But most of them will not be able to provide a ready supply of *chemicals* to make for plastics and other materials. This is why we must use our remaining fossil fuels wisely.

Nuclear power is one option, but what other sources of energy are available in the world around us?

Solar energy

Vast amounts of energy pour down onto the Earth from the Sun every day – a staggering 15 000 times as much as our technological society uses! The problem is, of course, that it is spread out over the entire 'daylight' face of the globe. It also tends to be at a peak where it is least needed – in the desert and at sea. Even so, it can be the equivalent of a 1 bar electric fire on every square metre of the surface. No wonder too much sun can burn!

How can we tap into all this 'free' energy? Many things use **solar cells** which produce electricity directly from sunlight. Examples are solar powered calculators and most types of spacecraft. Solar cells are expensive and, of course, are not very efficient if it's cloudy!

Solar furnaces use mirrors to concentrate the sunlight onto one spot. This can produce temperatures of up to 4000°C, which can be used to drive a generating plant, or directly as a heat source for industry.

Solar panels

A 'gentler' way of using solar energy involves **solar panels** to heat our homes. These use sunlight to heat water in long pipes on the roof. These are painted black to absorb the energy (see 4.8).

Even in cloudy Britain, this method could keep a house warm (and even provide hot water) for most of the year. In midwinter, it would need to be supplemented with other forms of heating, but it would still keep the fuel bills down!

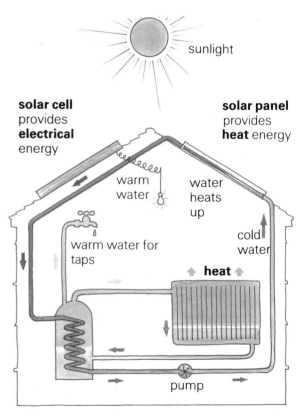

Geothermal energy

Another possible source of energy is the Earth itself! As you go deeper in the crust of the Earth, so the temperature rises, making life hot and sticky for deep miners. Now we have learned how to tap this **geothermal** energy. Water is pumped down to the hot rocks through boreholes, and there it boils. The steam produced is then collected and used to drive turbines and dynamos and so generate electricity. In some places, pipes are laid over cooling hot rocks near volcanoes. Water passed through the pipes gets very hot. It is pumped to nearby towns – providing free hot water!

Plant power!

Plants absorb energy from sunlight and lock it up in chemicals. Scientists in the USA are developing a type of cactus that produces an oily sap which can be refined like crude oil. In Brazil, oil is scarce but sugar cane grows in abundance. The sugar is fermented to make dilute alcohol. This is concentrated by distillation to give industrial alcohol. This is then used in cars as **'gasohol'**, instead of petrol!

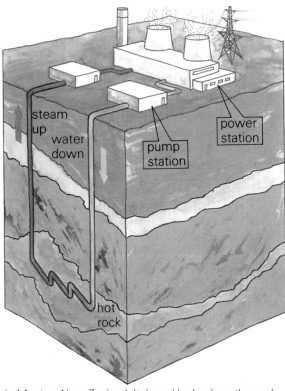

In Mexico, New Zealand, Italy and Iceland geothermal power stations like this are in use.

Using the weather

A large proportion of the solar energy heats up the air, causing it to move around as strong winds. The energy in wind can be seen by the scale of destruction left by a hurricane. The **wind** has also been used for centuries to grind corn and to pump water. Modern windmills use propellor-like blades which drive dynamos and so generate electricity. The weather can also provide energy from rain. The rain water forms rivers that can be trapped by dams. The rivers' potential energy is used to drive **hydroelectric** generators.

Energy from the sea

There are two ways of getting energy from the movement of the sea. **Wave power** uses the rocking motion of the waves to power dynamos. **Tidal power** involves trapping the high tide. The water is then made to flow back to the sea through turbines – which generate electricity.

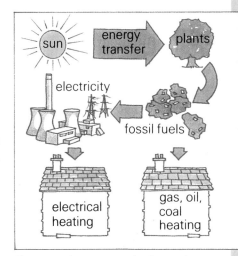

The more times energy is changed from one form to another, the more energy is wasted.

1. List *five* alternatives to fossil fuels. Include *one* that provides a useful chemical.

2. Why are solar cells not always a 'practical solution'?

3. How does a solar furnace work?

4. How can the weather be a source of energy?

5. Why is it more efficient to use fossil fuels to supply heat *directly* rather than using electricity?

4.14 Nuclear energy

Radioactive decay

Matter is the 'stuff' from which everything is made. Normally, matter and energy are quite separate and distinct – but this is not always the case. The smallest stable particles of matter are called **atoms**. In the case of radioactive material such as **uranium**, the atoms are unstable and can break down – **radioactive decay**. When this happens, a tiny fraction of the matter turns into energy – *a lot of energy*!

The famous scientist Einstein, showed just how much energy with his equation, $E = mc^2$. E is the *energy* released by a *mass m* in a nuclear reaction; c is the speed of light. As the speed of light is very high (300 million metres per second), the amount of energy that can be produced in this way is enormous. 1 unit of matter, if totally converted, will produce 90 000 000 000 000 000 units of energy! This breakdown of large, unstable atoms is called **nuclear fission**.

The international symbol for a radiation hazard.

Nuclear reactors

In a nuclear reactor, unstable 'nuclear fuels' (uranium and plutonium) are allowed to break down under very controlled conditions. The energy produced is used to boil water, make steam, turn a turbine and drive a generator to make electricity in the usual way. About 10% of British electricity is produced in this way – at slightly below the average cost.

Britain has had nuclear reactors in service for 25 years now. The early types needed a special kind of uranium which had to be separated from the 'raw' fuel. The rest of the uranium was then stockpiled. But nowadays modern reactors can use any uranium. By using the old stockpile, Britain already has enough nuclear fuel to produce five times as much energy as all of our North Sea oil!

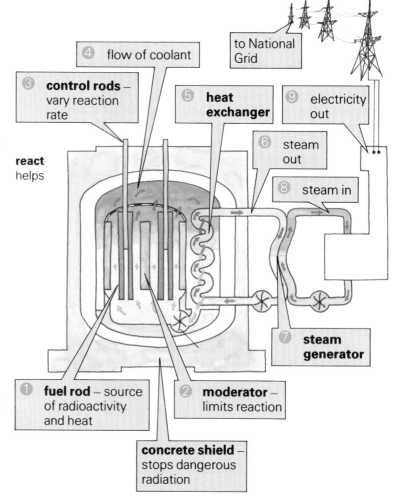

Nuclear problems

So why worry about fossil fuels if we have nuclear energy? Well, there are many reasons. One is that this same nuclear reaction is responsible for the horrors of nuclear weapons, such as that which destroyed Hiroshima. Countries using nuclear power stations might also learn to make **nuclear weapons** of their own, and the super-powers already have more than enough to destroy the planet – fifty times over!

Also, many people are worried about the nuclear plants themselves. Nuclear reactors produce powerful **radiation** which can kill in large doses. At lower levels, this radiation can also cause cancer and other problems. Because of this, the safety precautions are *very* thorough – much better than in a traditional plant. But human or mechanical error can never be entirely ruled out. There have been **radiation leaks** at nuclear plants all over the world. In the worst accident so far, a serious fire at the Chernobyl plant in Russia caused a very large leak of radioactive material.

The radiation energy from the atomic explosion can cause death and destruction. A nuclear power plant can control this energy to generate electricity.

Nuclear waste

Perhaps more worrying is the fact that nuclear reactors produce lots of radioactive waste. Some of this can be useful if it is reprocessed to make new fuel. But much of it has to be disposed of – and safely. Unfortunately, some radioactive wastes can take thousands of years to become safe, so what do we do with it? Bury it deep in mines? Dump it in the sea? Fire it into space? All of these methods have been investigated at various times.

Clean fusion

In the sun, a different type of nuclear reaction takes place. Small, simple atoms (hydrogen) are being *fused* into larger atoms (helium). This type of **nuclear fusion** is potentially a major step forward for nuclear power. Neither the starting material nor the waste is radioactive! Unfortunately, this reaction needs an enormous 'kick-start' of energy and scientists have not yet mastered this reaction. But work is in progress to harness this vast clean energy for the benefit of everyone. Perhaps nuclear fusion power will be the answer for the future? The problems of our waste-producing 'fission' reactors could be forgotten within a few decades – if we can survive that long...

1 What is unusual about the atoms in radioactive materials?

2
  a What happens to the 'lost matter' in nuclear reactions?
  b Why is so much energy produced?

3 List the problems caused by using nuclear fuels.

4
  a What happens in a **fusion** reaction?
  b Why is it safer than a **fission** reaction?

4.15 The nuclear debate

A tricky question

The nuclear debate has no easy answers. It is related to how we use energy now – and how we plan to use it in the future. The important thing is to understand exactly what both sides are talking about...

WE NEED IT!
Fossil fuels will run out. 'Alternative' energy sources are unproven on a large scale. We have large amounts of nuclear fuels already, and we know that it can provide all the energy we need. Once fusion is perfected, we will have limitless energy at our disposal!

STRICT LAWS
The laws are very strict about the levels of radiation allowable in nuclear plants. There is often more radiation in places such as Dartmoor or Aberdeen, which are built on granite.

DEMAND IS INCREASING
More and more energy is being used. People want a more comfortable lifestyle.

IT'S CHEAP
Nuclear reactors in Britain produce electricity more cheaply than 'traditional' power stations.

IT IS SAFE
Even including Chernobyl, there have been fewer accidents and injuries to workers in nuclear plants than in 'traditional' power stations. Look how dangerous coal mining is by comparison! What about the miners killed by accidents and by lung disease? The chance of a major accident at a nuclear power plant in Britain is tiny. Whenever problems have arisen, the built-in safety systems have always worked. Chernobyl was the results of a whole series of 'human errors' that could never have happened here. In fact, Chernobyl has so highlighted the dangers that it will probably never happen again anywhere in the world.

ITS ALL UNDER CONTROL
If all the electricity used by one person in their entire life was generated using nuclear fuels, the total amount of waste fuel would be about the size of a cricket ball! From a coal-fired plant, there would be tonnes of toxic chemical waste to dispose of – not to mention the nasty waste gases and the 'acid rain' they produce.

TERRORISTS COULD USE ANYTHING
A bucket of cyanide from the local factory thrown into the reservoir could kill thousands, too! Unfortunately, terrorists could use almost anything if they wanted to. It's something we have to live with, and try and guard against.

Organise a 'nuclear debate' in your class. Each side should prepare its case in advance. Use newspaper stories to support your ideas.

WE DON'T NEED IT!
The 'alternative' energy sources have not had as much money spent on their development as the nuclear reactors. If we conserve our fossil fuels, there will be plenty of time to perfect the new sources. And they do no harm to the environment. What's more, they will last forever – and are free!

DANGEROUS RAYS!
Nasty rays come from radioactive materials like nuclear fuel and waste. These can cause cancer and other long-term problems.

CUT DEMAND!
It's easy, just get people to insulate their homes. That would reduce demand.

NO IT'S NOT
The latest coal-fired power stations will be cheaper – and we've still got lots of coal left. Rivers, wind and tides can provide cheap *safe* power too!

IT IS DANGEROUS!
However good the system, there is always the possibility of human error, as Chernobyl showed only too clearly. It may not have gone up like an atomic bomb, but it spewed its radioactive waste into the atmosphere. It caused untold damage to the local environment. It also spread with the winds across frontiers and seas. It fell with the rain on Welsh hills, contaminating the sheep and making their meat inedible for many months. And who knows what the long-term effects will be? How many will die untimely deaths of cancer? Will the lives of children yet unborn be affected? The risks are just too great to take.

DANGEROUS WASTE!
The waste from nuclear reactors remains dangerous for hundreds of years or more. Storage problems are bad enough now, but if we went over to nuclear fuels altogether we would be buried in nuclear waste. There is much more than just waste fuel – there's disused equipment, tonnes upon tonnes of it, all radioactive. How could we leave such a menace to our children and our childrens' children?

TERRORISTS COULD USE IT
Nuclear waste contains plutonium. This is also used to make atomic bombs! What is there to stop a terrorist group getting hold of this and making a bomb of their own? Or spreading it around the cities to make them uninhabitable?

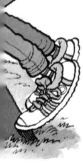

MODULE 4 ENERGY

Index
(refers to spread numbers)

A
absorbers 4.8
alternative energy sources 4.10
anthracite 4.11
atomic explosion 4.14
atoms 4.14

B
batteries 4.1, 4.2
biotechnology 4.10
bituminous coal 4.4.11

C
carbon 4.12
cavity walls 4.8
Celsius 4.6
centigrade 4.6
central heating 4.4.9
charcoal 4.11
coal 4.9, 4.10, 4.11, 4.12
combustion 4.12
conduction 4.8
conservation of energy 4.10
contraction 4.6
convection 4.8

D
draught excluders 4.8, 4.9
dynamos 4.5

E
electricity 4.3, 4.4, 4.5
energy
  active 4.2
  chemical 4.3
  electrical 4.1, 4.2, 4.3, 4.5
  electromagnetic 4.1, 4.3
  geothermal 4.13
  heat 4.1, 4.2, 4.3, 4.6
  kinetic 4.1, 4.3, 4.4, 4.5
  light 4.1, 4.2, 4.3
  mechanical 4.2
  nuclear 4.10, 4.14
  potential 4.1, 4.2, 4.4, 4.5
  solar 4.13
  sound 4.1, 4.3
  stored 4.1, 4.2
expansion 4.6

F
fibreglass 4.8
fission (nuclear) 4.14
fixed points (thermometers) 4.6
fluorescent light 4.3
foam 4.8
food 4.11
fuels 4.2 (fossil) 4.10

G
gas (as fuel) 4.9, 4.11, 4.12
gasohol 4.13
generating 4.3
generating hall 4.5
geothermal power 4.13
glucose 4.1

H
heat flow 4.6
heating 4.9
hydrocarbons 4.12
hydroelectric power 4.13
hydrogen 4.12
hypothermia 4.6

I
immersion heater 4.9
industrial fuels 4.10
insulation 4.8, 4.9
insulators 4.8, 4.9

L
lignite 4.14
LPG 4.9

M
matter 4.13
microwaves 4.9

N
National Grid 4.4
natural gas 4.13
night storage heaters 4.4
North Sea oil 4.10

nuclear debate 4.15
  fission, fusion 4.14
  waste, weapons 4.14

O
oil 4.9, 4.10, 4.11, 4.12
OPEC 4.10

P
peat 4.11
petrol 4.1, 4.
plants 4.11
pollution 4.9, 4.12, 4.14
power 4.3, 4.4
pumped storage system 4.5
purification 4.13

R
radiators 4.8
roof insulation 4.9

S
smokeless zones 4.11
solar cells/panels 4.13
sulphur dioxide 4.12

T
temperature 4.6
thermometers 4.6
tidal power 4.13
transmission unit 4.5
turbines 4.4, 4.5, 4.14

V
vacuum flask 4.8

W
waste
  heat 4.3
  nuclear 4.14
watt 4.3
wave & wind power 4.13
wood 4.11

Photo Acknowledgements

The references indicate the spread number and, where appropriate, the photo sequence.

Barnaby's (M P Brown) *4.9/2*, (Malcolm Pendrill) *4.10*; British coal *4.11*; J Allan Cash, contents, *4.2/1, 4.3/1, 4.11*) Sally & Richard Greenhill *4.2/2, 4.3/2*; GeoScience Features *4.12, 4.1/2*; Trevor J Hill *4.1, 4.2/2, 4.3/3, 4.6, 4.9/1*; Frank Lane Agency *4.8*; Rex Features 4.10; North Scotland Hydroelectric Board *4.4*; Science Photo Library (Alex Bartel) *4.6*; US Navy *4.14/1*; Sporting Pictures *4.8*; Zeithoper Photoreport *4.10*

Picture research: Jennifer Johnson

MODULAR SCIENCE
for GCSE

MODULE 5 *Materials*

*Different **materials** have different properties. They behave differently under certain physical and chemical conditions. These differences help us to use and develop materials for particular purposes. This module will help you understand more about the nature of materials so that you can recognise, understand their properties and classify them.*

Relevant National Curriculum Attainment Targets: (6), (7), (8), (10)

5.1	Making use of materials
5.2	Shaping up
5.3	Look and feel
5.4	Conductors and in
5.5	Heavyweights and lightweights
5.6	Strength
5.7	Composite materials
5.8	Hard and hardwearing
5.9	Rotten materials
5.10	Change of state
5.11	Crude oil
5.12	Plastics
5.13	Flammable materials
5.14	Materials – 'tailor-made' for your use!

Index/Acknowledgements

MODULE 5

5.1 Making use of materials

Everybody does it!

Humans are different to all the other animals in many, many ways. One way is our ability to make use of a wide variety of materials. Even our Stone Age ancestors used lots of different materials – although they mainly used only natural materials that they could find all around them.

Shelters were originally in caves in the rocks.

Plant juices were used to paint on walls.

The skin or fur of animals was used for clothing.

Bone and wood were strapped together with leather to make spears.

Flint was used to make other tools such as spears and axes.

Simple stone tools made from large pebbles were used to cut meat.

There were not too many people around at the time, and they used only a few materials. Most of these materials came readily to hand. If only things were so simple now...

Everyday materials – everywhere!

To-day we still use natural materials, such as wood, but many of the materials that you use everyday are made by us. These 'man-made' or synthetic materials, such as plastic, have to be made in a particular way from natural materials. One advantage of synthetic materials is that they can be made to suit a particular use. Look at the picture below and try to think of reasons why we use the different materials shown.

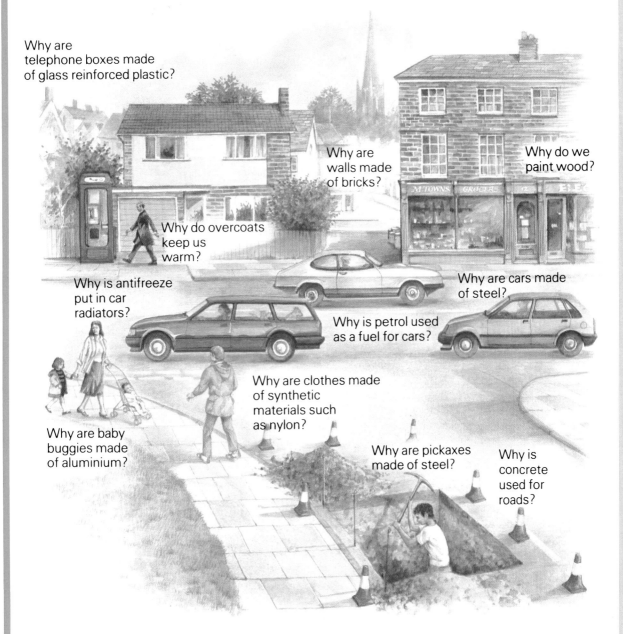

To-day there are many more people using a lot more materials with many different properties. These properties make the materials very useful. But exactly what *are* the properties of materials? How are materials made? What will happen if we use up all the materials?

To find out more, read on ▶

5.2 Shaping up

Do shapes last?

You have to apply a force in order to shape materials. This force can be such things as a pull or a stretch, a push or a squeeze. If you stretch a rubber balloon by blowing it up, it returns to its original shape when the air is let out. It is said to be **elastic**. Other materials such as pottery clay take on new shapes. These materials are said to be **plastic**.

Clay can be moulded to stay in a new shape.

Let go and the balloon will return to its original shape!

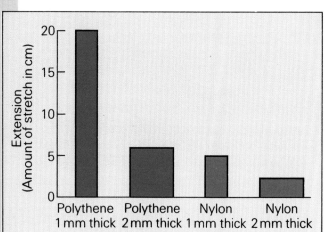

Stretching . . .

Look at the graph. What does it show you about the stretchiness of polythene and nylon? The stretchiness of materials depends not only on the *type* of material but also its *thickness*. What does the graph show you about the change in length of the polythene as it becomes thicker? As the thickness of a material increases, it becomes harder to stretch.

. . . to the limit

Look at the graph. What does it show you about how the length of an elastic material increases as you increase the pull? When a material is elastic, if you double the pull you double the increase in length and so on. However, as the pull increases, the material takes on a new shape. It no longer returns to its original shape. Increasing the pull even more makes the material go more out of shape and eventually it breaks.

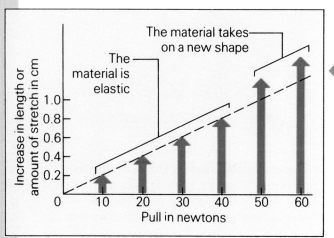

Materials will stretch only so far before they take on a permanent new shape.

The shape of things

When a material no longer returns to its original shape, it has been permanently stretched. This permanent stretching is called **plastic deformation**. Materials which are capable of large plastic deformations can be shaped more easily. They are said to be **ductile**. Some plastics are ductile so they can be easily shaped into useful objects for the home.

The materials used to make these things were easy to shape – they are called 'plastics'.

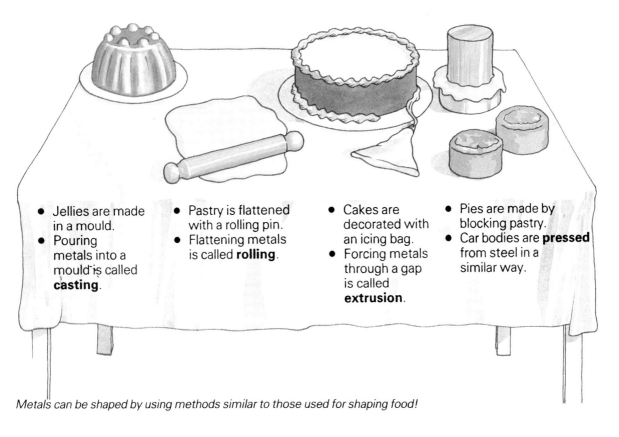

- Jellies are made in a mould.
- Pouring metals into a mould is called **casting**.

- Pastry is flattened with a rolling pin.
- Flattening metals is called **rolling**.

- Cakes are decorated with an icing bag.
- Forcing metals through a gap is called **extrusion**.

- Pies are made by blocking pastry.
- Car bodies are **pressed** from steel in a similar way.

Metals can be shaped by using methods similar to those used for shaping food!

1 Give one example, in each case, of an elastic and plastic material.

2 What is the difference between an elastic and plastic material?

3 State three things which affect the stretchiness of materials.

4 Why do thin woollen jumpers lose their shape more easily than thick ones?

5 Look at the bottom graph opposite.
  a What is the largest pull, in newtons, the material can be given and still remain elastic?
  b What is the smallest pull, in newtons, the material can be given to take on a new shape?

6 Metals are ductile – they can be easily stretched into new shapes. Give two examples of shaping metals, using this property.

5.3 Look and feel

Looks are Important

The different appearance of materials can be put to good use.

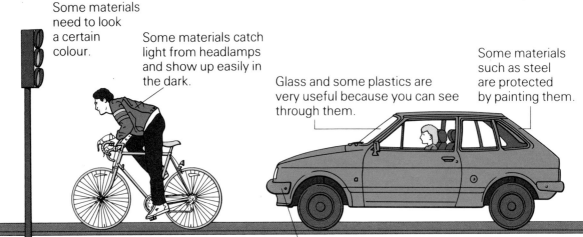

Some materials need to look a certain colour.

Some materials catch light from headlamps and show up easily in the dark.

Glass and some plastics are very useful because you can see through them.

Some materials such as steel are protected by painting them.

Chrome protects metals which are scratched too often to use paint.

Looking the same

All materials are made up of tiny particles. The appearance of a material depends on the type of particle it is made from and the way these are arranged. Metals, for example, are made up of particles called **atoms** arranged in a similar way so that most metals look very similar. Different plastics, such as polythene and polypropylene which are used in packaging food, are difficult to tell apart because they have the same atoms arranged in a very similar way.

A rough diamond looks like a piece of glass because they have a similar arrangement of particles.

Looking different

Some materials can be made to look different by adding other materials. Glass and plastic, for example, have other materials added to give them colour. Some materials have their appearance altered by coating their surface. Wood can be painted to make it a different colour or varnished to make it look shiny. Metals, such as iron, have a coating of another metal, for example chromium, put on them by a process known as **electroplating.**

Which materials in this picture have had their appearance changed?

Feeling right

The 'feel' of a material is called its **texture**. Some materials are made to have a smooth texture, others are made with a rough texture. It all depends on what they are to be used for . . .

Some clothes such as wool have a soft texture.

Sharp metals may have to be smoothed down.

Plastics feel smooth.

Abrasives such as 'wet and dry' must feel rough.

Tyres must feel rough if they are to get a good grip on a road surface.

The large crystals of zinc make zinc-coated metals feel rough.

Feeling different

The texture also depends on the type of particle in a material and the way they are arranged. Some materials have a regular arrangement of particles called a **crystal**. These crystals can join up to form grains of different sizes. The texture depends on the size and hardness of the grains. If the grains are large and hard the material feels rough. **Abrasives**, such as sandpaper, are used to make the grains on the surface of a material, smaller. Some materials, such as wool and plastics are not made of crystals – they have a different structure.

1. What is the name given to a material you can see through?

2. Why do some plastics have a similar appearance?

3. Name four articles which are often chromium plated.

4. If you wanted to coat a nickel wedding ring with silver what process would you use?

5. Why are smooth tyres on cars dangerous?

6. **a** Why does the surface of a metal normally feel rough?
   b What does an abrasive do to the surface?

5.4 Conductors and insulators

Energy on the move

Different materials can be used to control the movement of energy...

- Electricity is carried along the wire to the kettle.
- Heat passes through the pan to the water.
- Heat from the pan cannot pass along the handle.
- Electricity cannot pass through the outside covering of the wire.
- Heat from the kitchen cannot pass into the fridge.

Heat and electricity are useful forms of energy – but to make full use of them, you have to control their movement.

Conductors and insulators

Materials which allow energy to pass along them easily are called **conductors**. Those which do not are called **insulators**. Metals such as silver and copper, are good conductors of heat and electricity. Materials like wood, rubber and plastics are all insulators. They do not allow energy to pass along them as easily as conductors.

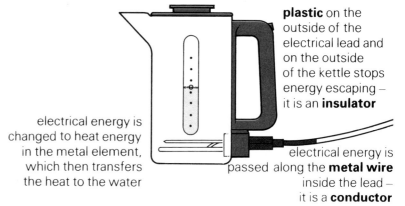

plastic on the outside of the electrical lead and on the outside of the kettle stops energy escaping – it is an **insulator**

electrical energy is changed to heat energy in the metal element, which then transfers the heat to the water

electrical energy is passed along the **metal wire** inside the lead – it is a **conductor**

Insulators and conductors are often used side by side.

Spoilt for choice

All metals are conductors but some are better conductors than others. Look at the picture. Along which metal has the heat travelled furthest in one minute? Silver is a better conductor than either iron or copper but it is very expensive ... so it is not a good choice for everyday use.

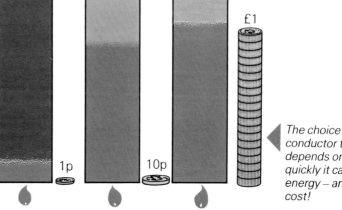

IRON — 1p
COPPER — 10p
SILVER — £1

The choice of which conductor to use depends on how quickly it can transfer energy – and its cost!

Saving heat

Insulators are useful for stopping things getting too hot. They stop heat energy from escaping. Metal saucepans have plastic or wooden handles to stop them from burning your hands. Air is a good heat insulator, so anything that traps air is also a good insulator. This is why glass fibre is used for loft insulation. It costs a lot to make heat energy, so if you let heat escape, you are wasting money ... it is worth insulating your home against this waste.

Clothes and fur keep us and animals warm by trapping air in layers. This air keeps the heat in.

This electrical equipment is well insulated by plastic coated wire and plastic drill casing. Even the ear 'mufflers' insulate against sound energy!

Playing safe

Electricity, carrying a lot of energy, can harm or even kill you. To stop this happening all electrical plugs, sockets and leads are made with thick plastic covers. Tools and other electrical appliances, such as hairdriers, have insulated rubber or plastic handles. Some insulators can become conductors, if they are carrying a lot of energy. Water, for example, will conduct electricity from the mains supply. This is why electrical appliances should never be used near a bath full of water.

1. Name two forms of energy.

2. Name three materials which are good insulators.

3. Look at the metal rods opposite. If you held each metal at the top end, which metal would feel warm first?

4. Why are metals conductors?

5. **a** Animals have fur or feathers to keep warm. What clothes do you wear to keep warm?
   b Explain how these clothes keep you warm.

6. Why do TVs have warning signs at the back of them?

5.5 Heavyweights and lightweights

A dense situation

You can compare the heaviness of two materials by putting them at each end of a see-saw. You can also compare the **density** of two materials in this way. This is a measure of how much 'stuff' of a material is packed into a particular volume of it. Look at these pictures of two different materials on a see-saw...

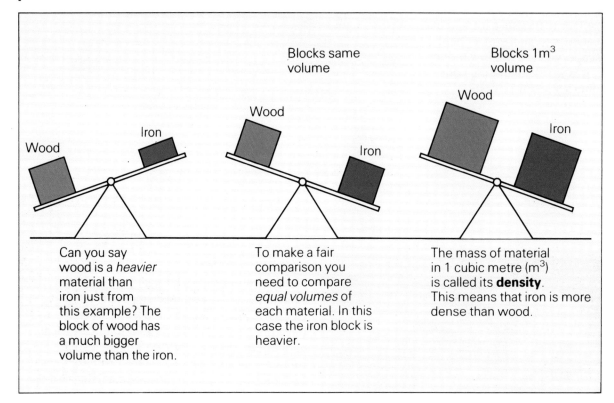

Can you say wood is a *heavier* material than iron just from this example? The block of wood has a much bigger volume than the iron.

To make a fair comparison you need to compare *equal volumes* of each material. In this case the iron block is heavier.

The mass of material in 1 cubic metre (m^3) is called its **density**. This means that iron is more dense than wood.

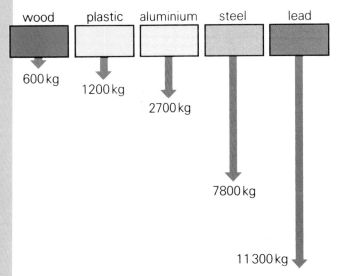

Some materials, such as lead, are more tightly packed than others.

What makes materials 'heavy'?

Some materials, such as metals, are neatly packed and close, together, like biscuits in a packet. Other materials, such as plastics, are not so neatly packed and have large gaps of air between them. Look at the diagram which shows the masses of one cubic metre (m^3) of different materials. What does it show you about how the type of packing of the materials affects their density?

'Light' is right

'Lighter', or less dense, materials need less energy to lift or carry them. Bricks are about three times more dense than wooden planks. A person working on a building site, would use three times more energy lifting a hodful of bricks than lifting wooden beams of the same volume... but maybe a handful of bricks is slightly easier to handle!

Brick is more dense than wood... a hodful of bricks is quite a heavyweight compared to a plank of wood!

Changing density

Some objects that are not dense enough can be made denser by filling up the air gaps with denser materials. The steel hull of a submarine or the hollow plastic base of a sunshade can be made denser by replacing the air with water. Materials that need to be made less dense (lighter, but with the same volume), can have air bubbled into them. For example, polystyrene has air blown into it to make much lighter **expanded polystyrene.**

A lightweight shade can get blown over. Fill the base with water and it becomes denser, and so heavier and more stable.

'Heavy' hits back

It might seem that 'heavy', dense materials have little use. However, walk around any town or glance around the school or home and you will see examples of objects such as litter bins, paper weights and laboratory stands that are heavy enough to stay in one place but not so heavy they cannot be lifted. And some things, such as statues and paving slabs, shouldn't move at all!

1. How many times more dense than plastic is wood?

2. Explain why lead is more dense than plastic.

3. Why are aeroplanes made of aluminium and not steel?

4. A man, lifting bricks, used '15 packets of cornflakes' worth of energy. How much would have been used lifting the same amount of wood?

5. Explain why wet clothes are heavier than dry.

6. Why is packaging material often made of expanded polystyrene?

5.6 Strength

What is strength?

A strong material is one which is difficult to break when you apply a **force**. This force can be a pull such as a climber would use to test a climbing rope. It can also be a squeeze or a crushing blow such as a builder might use on stone slabs when making crazy paving. A material which is difficult to break by pulling is said to have good **tensile strength**. One which is difficult to break by crushing is said to have good **compressive strength**.

This climber is glad that a thick nylon rope has good tensile strength.

This job is hard work because concrete has good compressive strength.

Strong stuff

Some materials are used because they have good compressive strength. A good example is a brick which has to be able to withstand the squashing of all the other bricks above it. It is about 20 times more difficult to squash a brick than to stretch it.

The usefulness of ropes, lines and chains depends on their tensile strength. Look at the bar chart. What does it show you about the increasing use of synthetic fibres such as nylon for ropes instead of natural ones such as manilla or sisal?

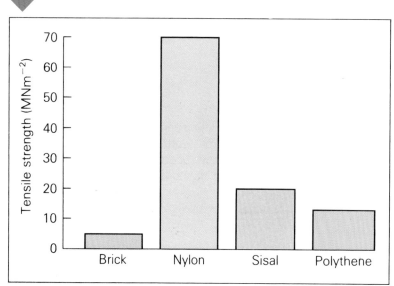

Tensile strength (MNm^{-2}): Brick, Nylon, Sisal, Polythene

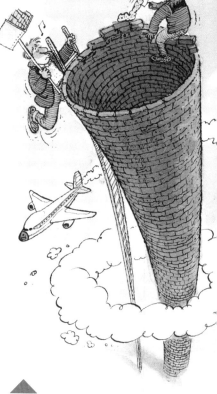

One brick can support the weight of 40 000 others before it will be crushed.

Strength and area

The tensile strength of a material depends on its cross-sectional area. Look at the diagram below. If the area of the material is 4 times larger, the force required to break it needs to be four times as big. The compressive strength can be increased in the same way.

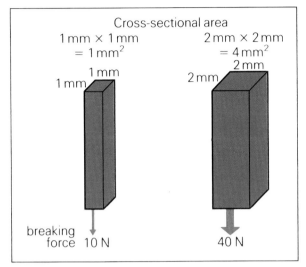

By increasing the thickness four times, a material will become four times stronger.

Very thin spider's webs still have great tensile strength. If 200 times thicker, their silk would support a car!

Bending

A material which is being pulled, is made longer and it is said to be under **tension**. A material which is being squashed, is made shorter, and it is said to be under **compression**. If the opposite sides of a material are being squashed and stretched at the same time, it is being **bent**. If a material is easy to bend it is **flexible** and it has good compressive and tensile strength.

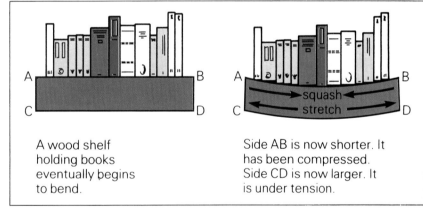

A wood shelf holding books eventually begins to bend.

Side AB is now shorter. It has been compressed. Side CD is now larger. It is under tension.

If a shelf is to resist bending, the material must have a high tensile strength **and** a high compressive strength.

1 Give one example of a material which has good **a** compressive strength, **b** tensile strength.

2 Look at the bar chart.
 a What is the compressive strength of a brick if it is 20 times greater than its tensile strength?
 b What is the tensile strength of nylon?
 c Why do climbers prefer nylon ropes to natural fibres such as sisal?

3 Look at figure at the top left of this page. What would be the breaking force if the cross-sectional area were 100 mm^2?

4 The tensile strength of steel is about 2100 Nm^2 and that of nylon about 70 Nm^2. If a steel wire had a cross-sectional area of 10 mm^2, what would the area of nylon rope need to be in order to have the same tensile strength?

5 Concrete has good compressive but poor tensile strength. Why is it not flexible?

5.7 Composite materials

Tough and brittle

Some useful materials such as glass and concrete are **brittle**. A brittle material usually cracks under a force and any more energy being used to break it passes along the crack making it bigger.

Tough materials such as steel used for car bodies are not as easy to break. If a force is applied to a tough material (such as a car tyre), it absorbs energy from the force. A tough material often absorbs energy by changing shape slightly – but it will not break.

Although tough materials and brittle materials have separate uses, sometimes they may be combined to make more useful materials.

Concrete is brittle; under large forces it will crack rather than gradually change its shape.

Making materials tougher

Materials can be made tougher if you can stop them cracking. In order to do this, brittle materials such as plaster and some plastics are combined with a material made of fibre such as glass fibre or paper. The energy used which would normally break the brittle material is passed along the fibre instead. This principle is used to make glass reinforced plastic (GRP) which contains glass fibre in a brittle plastic resin. Such materials are called **composite materials**.

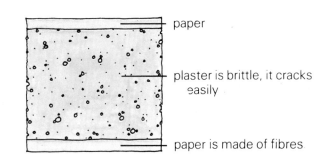

Plasterboard is made by coating a plaster core with paper fibres. The paper helps to stop the plaster from cracking.

The bottom section of the bridge is being stretched by the weight of the cars passing over it. Concrete is brittle when it is stretched and can crack.

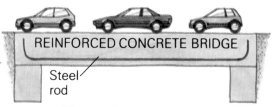

The steel rods act like fibres. Energy being used to try and crack the concrete is passed along the steel rods instead.

Bridges of only concrete crack under the weight of cars. Steel rods prevent the cracks from forming.

Making materials stronger

Some materials, such as wood, are strong in certain directions. Plywood consists of layers of wood bonded together. In one layer the grains run *across* the wood, in the next layer the grains run *along* the wood. Look at the picture. What does it show you about how more strength can be given to the wood by making it plywood?

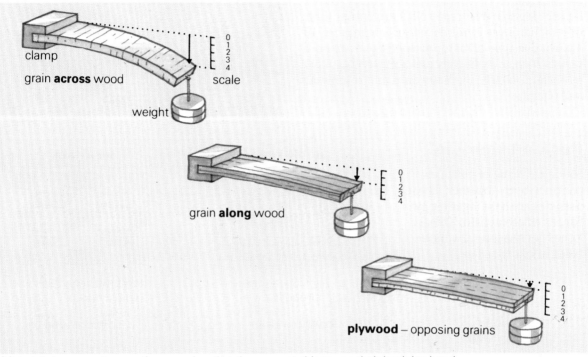

Materials become stronger if arranged so that they are not able to use their 'grain' to bend.

Materials can also be made stronger by altering their shape. Large thin sheets of plastic are too bendy to use as roofing material. Cardboard boxes, used in packaging, also need to be stiff to protect their contents. One way of improving the stiffness of materials is to alter its shape by **corrugating**.

Materials break when forces act on a weak point. Corrugation helps to spread these forces around, making the material stronger. Folded newspaper can support a heavy weight of water!

1 Name one brittle and one tough material.

2 What is the difference between a brittle and a tough material?

3 Explain why horsehair used to be added to plaster.

4 Look at the diagram of wood, above.
  a Is wood stronger when weights are placed along or across the grain?
  b In what direction do you think wood grain should run in a beam supporting a roof?

5 Name two advantages of using corrugated cardboard instead of cardboard sheets for packaging.

5.8 Hard and hardwearing

What is hardness?

A **hard** material is one which is difficult to scratch or dent. It is a property which only solid materials have. All solids are hard compared to liquids or gases but some solids are harder than others. Hardness is not necessarily a property of a particular type of material since there are hard woods such as mahogany and soft woods such as balsa. Similarly, there are hard and soft metals and plastics.

Synthetic diamonds are extremely hard

Packing them in!

Solids are made of particles packed closely together. These particles are held together by **forces of attraction.** These forces act like mortar which holds bricks together.

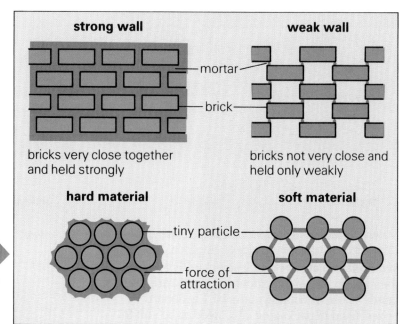

Most hard materials have very closely packed particles held together by strong forces. Soft materials are usually less closely packed and held only weakly.

The hard stuff

Tungsten carbide is a very hard metal alloy and is used to drill through steel or concrete. Metal cutters made of steel can slice through tin. The rubbing action of sandpaper cuts into wood, producing a smooth finish. Look at the table. What does it show you about how the hardness of the materials is related to their use?

Hardness index	Material
10	Diamond
9·7	Silicon carbide
8·5	Tungsten carbide drill
7→5	Steel
7·0	Sand
5·5	Glass
5·5	Nickel
5→4	Concrete
1·5	Tin
3→1	Wood

The harder a material, the higher its number in the hardness index.

Hardwearing

Materials can wear out from the constant rubbing action of other materials. Rubber bicycle brakes wear out and need replacing because they are worn out by the hard metal rims of the bicycle wheel. Materials that are hard, such as diamond, are not worn out by softer materials so they are **hardwearing.** Natural materials such as cotton from plants, and wool from animals, are not hardwearing. This is because in their natural state, even if they wear away, they can be replaced by the growth of new material.

If a sheep's coat needs replacing, new wool grows. If the clippers wear out, they have to be sharpened.

A longer life

Materials can be arranged so that they resist the rubbing action of other materials better. One way of doing this is to **pack** the material more tightly, for example, using a steam roller to make a road more hardwearing. Or in the case of a fibre, such as cotton, it can be woven more closely together.

Materials can also be **mixed** with other materials that are more hardwearing. Woolen jumpers often contain an artificial fibre (such as nylon) which is more hardwearing. This helps these jumpers to last through more wear and washing.

Made at first as tough clothes for working in, the material for hardwearing jeans is made by weaving tightly packed thick cotton fibres. ▶

1. Name one hard and one soft wood.

2. Explain why high impact polystyrene is harder than expanded polystyrene.

3. Look at the table on hardness.
   a Name one hard and one soft metal
   b Why is silicon carbide paper used instead of sandpaper to rub down car bodies?
   c Why is diamond used widely in industry as a cutting tool?

4. Why do the elbows of woolen jumpers wear out first?

5. Why does weaving a fibre more closely together make it more hardwearing?

6. Look at the labels on some of your clothes. What artificial fibres have been mixed with natural fibres like cotton and wool? List the combinations you can find

5.9 Rotten materials

Its rotten!

Materials that rot are being eaten away by **decomposers** such as bacteria and fungi. These are tiny organisms that are all around us. They are so small that you can't see them, and so light they are easily carried by the wind from place to place. When they land on food, they multiply or reproduce at a very quick rate. You can see the rapid growth of fungi as mould on food.

Air raid! Fungi and bacteria from the air grow rapidly once they find material to feed on.

What type of materials rot?

Fungi and bacteria, unlike plants, can't make their own food. They feed on the carbohydrates found in dead material such as paper and wood. They also need **damp** conditions but will eat any dead wood, whether in a forest or part of a house. When the wood is being eaten away, it loses all its strength. This can be a serious problem because wood is used in houses as a support in beams, window frames and floorboards.

Types of rot

Different fungi produce different types of rot but in this country they are divided into two types: wet and dry rot. Wet rot is fairly common but is easily cured. Dry rot is less common but more of a problem – it can lie 'hidden' in the wood and so may reappear long after treatment. Look at the pictures of wet and dry rot. How do you think they get their names?

A picnic in the woods for decomposers – rotted wood lies all around the base of this dead tree.

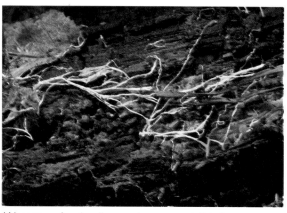
Wet rot – after landing on damp wood, this fungus produces white stalks (hyphae) which spread across the wood.

Dry rot – a tiny fungus which is difficult to control. It can even move along brickwork to attack nearby wood.

Prevention is better than cure

Dampness can enter buildings in a number of ways. This means conditions will then exist for wood to decay. New buildings should be designed using the right materials to prevent the possibility of damp getting in. Existing buildings must be properly maintained to keep damp out. Look at the picture. What does it show you about some of the causes of dampness entering a building? What must be done, do you think, to prevent the dampness?

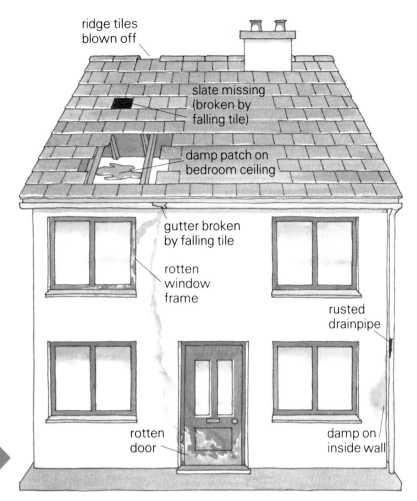

If the slate and gutter were repaired soon after the ridge tiles blew off, there would have been no other problems.

Breaking down, building up

Rotting is nature's way of breaking animal and plant material down so that it can be used again – **recycling** materials. If rotting didn't take place, then essential food for plants would soon run out. Gardeners make use of this process when they build a **compost** heap. However materials such as plastics, which don't rot are used today. Look at the bar chart. What does it show you about the use of plastics and paper compared with thirty years ago?

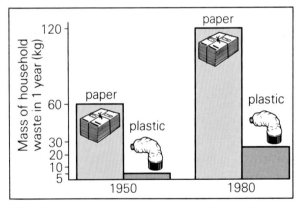

Nowadays we are making even more of a mess!

1 Why is it very difficult to kill all fungi?

2 State two conditions needed for fungi to grow and multiply.

3 Why is wood rot a problem and why is dry rot more serious than wet rot?

4 Look at the picture of the house. Identify four ways that dampness could get inside. Suggest possible cures.

5 Look at the bar chart.
   a By how much has the amount of plastics used increased in the last thirty years?
   b Why is 'plastic rubbish' a problem and how might it be solved?

5.10 Change of state

How do materials move?

Have you ever wondered how you can soon smell the gas when you've left the cooker on, but unlit? If you left a tap running, you would not know until much later, when the water flowed under your feet. Can you explain why snow piled up against a door stays in place, even when the door is opened?

These materials behave in different ways because they are in different **states** – one is a **gas**, one a **liquid** and the other a **solid**.

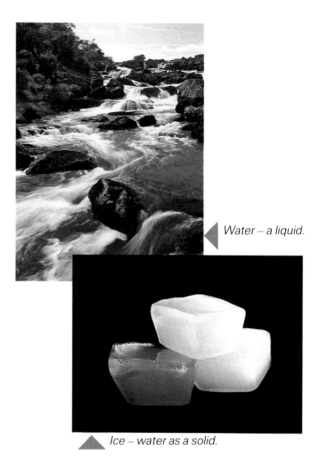

Water – a liquid.

Steam – water as a gas.

Ice – water as a solid.

Energy for a change

The faster the particles in a material are moving, the more energy it has. When a solid is **heated**, the particles **gain** energy and vibrate more. As they gain more energy, they are able to move much more freely and become liquids. By cooling a liquid, energy is removed, so the particles slow down and become solids again.

Altered states – solids to liquids . . .

The temperature at which a solid changes to a liquid is called its **melting point**. A material with a low melting point, such as nylon, melts more easily than one with a high melting point, such as iron. Materials can be made to melt more easily by adding other substances. Salt is added to icy roads in winter to make the ice melt more easily. Look at the graph. What does it show you about how the melting point of lead and tin can be changed by mixing them together?

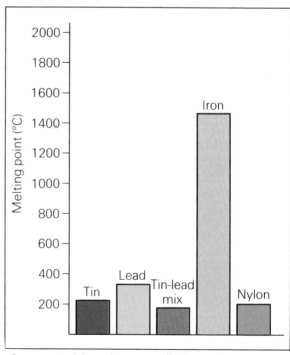

Some materials melt more easily than others.

...liquids to gases

The temperature at which a liquid changes to a gas is called its **boiling point**. Some liquids, like water, are made up of particles that are very small and light. These particles are free to move about easily, so only a little energy is needed to change them into gases. Such liquids have low boiling points.

Other liquids are made up of large, heavy molecules which need a lot of energy to make them move fast enough to become gases. These liquids have high boiling points, for example, cooking oil boils at 250°C.

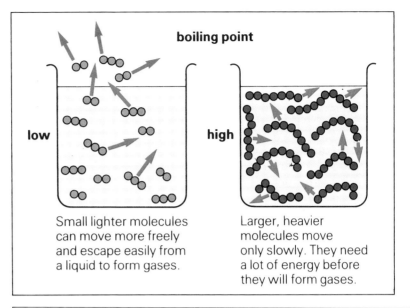

Small lighter molecules can move more freely and escape easily from a liquid to form gases.

Larger, heavier molecules move only slowly. They need a lot of energy before they will form gases.

All mixed up

Some liquids are made up of mixtures of different particles. Crude oil is an example of such a liquid mixture. Some of the particles are small and light; these particles boil before the heavier ones. By slowly increasing the temperature, first the light particles boil off – these can be cooled and **condensed** (collected as a liquid). After more heating, the heavier particles boil off – these too can be cooled and collected.

In this way, the mixture has been separated by a process called **fractional distillation**.

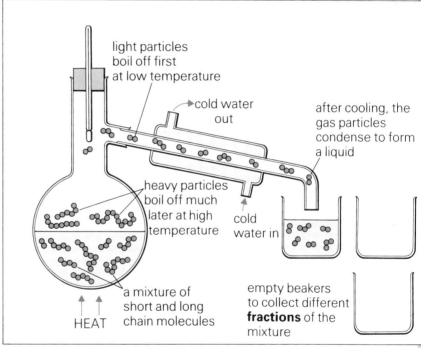

A typical apparatus for the fractional distillation of a mixture of liquids.

1 How does the movement in gases compare to that in liquids?

2 Why does steam turn back into water when it hits a cold window?

3 Look at the bar chart opposite.
   a How does the melting point of the tin–lead mixture compare with that of pure tin and pure lead?
   b Why is tin not used to make the wires of light bulbs which reach 800°C?

4 Why is oil more runny when it is hot than cold?

5 Why is antifreeze added to car radiators in winter?

6 Look at diagram of the boiling mixture.
   a Why is the mixture heated?
   b What is the purpose of the cold water?
   c How does the boiling point of the short molecules compare with that of the long molecules?

5.11 Crude oil

Where does oil come from?

The oil we now rely on for many of the materials we use today was formed many millions of years ago...

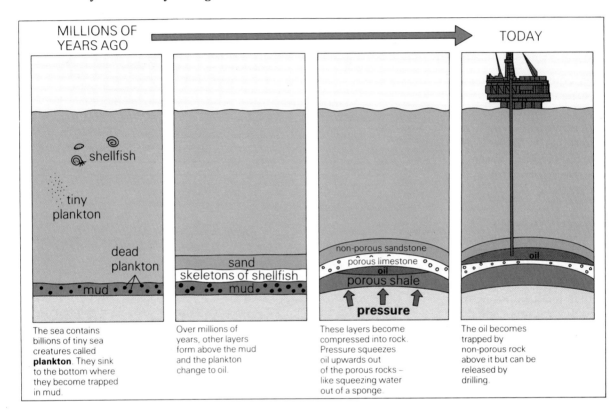

The sea contains billions of tiny sea creatures called **plankton**. They sink to the bottom where they become trapped in mud.

Over millions of years, other layers form above the mud and the plankton change to oil.

These layers become compressed into rock. Pressure squeezes oil upwards out of the porous rocks – like squeezing water out of a sponge.

The oil becomes trapped by non-porous rock above it but can be released by drilling.

What is oil used for?

Nowadays many materials with useful properties are obtained or made from crude oil like those shown in these pictures...

Nylon – a plastic which makes strong ropes, smooth stockings and hardwearing gears.

Plastics – these can be easily shaped and coloured.

Paraffin – burnt to give light and heat.

Solvents – such as nail varnish, and 'white spirit' for dissolving paints.

What is crude oil?

Crude oil is a mixture of compounds called **hydrocarbons**. Hydrocarbons contain particles called *molecules*. These molecules are made up of even smaller particles – *atoms* – of hydrogen and carbon joined together in chains. These molecules have a backbone of carbon atoms. The number of carbon atoms or the chain length is different for each type of hydrocarbon. Molecules with a long chain length are heavier and less runny than those with a short chain length.

Separating the mixture

As you have seen, crude oil is a mixture of many useful substances, all of which have different boiling points. These can be separated out by a process called fractional distillation (see 5.10). The oil is heated in a large column called a **fractionating column** where it is separated into the different parts, called **fractions**. This is done on a very large scale in the industrial process of the fractional distillation of oil.

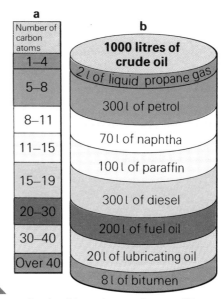

Figure 1. Crude oil is a mixture of many different parts. Each part of the mixture is made up of hydrocarbons with a particular number of carbon atoms.

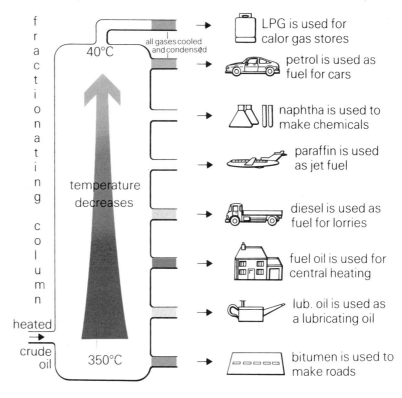

Figure 2. You can see how useful crude oil is once it is separated into its various fractions.

1. What is oil made from?

2. Look at the pictures of different materials on the opposite page. Which materials are obtained *directly* by fractional distillation of crude oil?

3. Look at Figure 1b above. Put the data shown in the form of a bar chart.

4. Look at Figure 1a. How does it show you that paraffin is a mixture of hydrocarbons and not a pure compound?

5. Look at Figure 1a. How does it show you that petrol is more runny than bitumen?

6. Look at Figures 1 and 2. What do they show you about the boiling point of a hydrocarbon as its chain length increases?

5.12 Plastics

Yet more uses of oil

Crude oil can be processed even more to produce 'extra helpings' of the parts that are in great demand. Other products can be made which are used to make **plastics** – materials which you probably use every day, all the time...

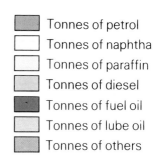

- Tonnes of petrol
- Tonnes of naphtha
- Tonnes of paraffin
- Tonnes of diesel
- Tonnes of fuel oil
- Tonnes of lube oil
- Tonnes of others

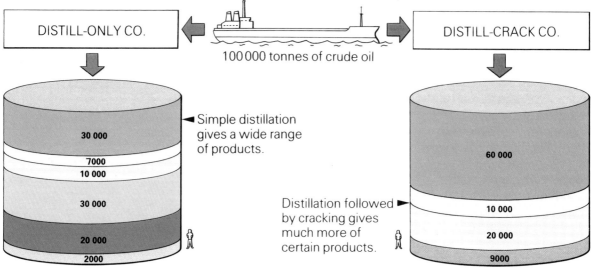

100 000 tonnes of crude oil

DISTILL-ONLY CO.: 30 000 / 7000 / 10 000 / 30 000 / 20 000 / 2000

Simple distillation gives a wide range of products.

Distillation followed by cracking gives much more of certain products.

DISTILL-CRACK CO.: 60 000 / 10 000 / 20 000 / 9000

If petrol, naphtha and paraffin are in great demand, which company will find it easiest to sell its products?

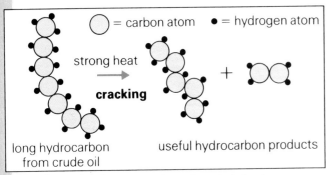

Strong heating causes the carbon chain in a long hydrocarbon to break – producing smaller hydrocarbons.

The fractions are made up of compounds called hydrocarbons (see 5.11). **Hydrocarbons** are molecules made up of hydrogen and carbon atoms joined together in a chain. The backbone of the chain consists of carbon atoms. Heating the molecule causes it to vibrate more. Continued heating will vibrate the molecule enough to break the carbon chain. This means that long chain molecules can be shortened. The breaking of the chain is called **cracking**.

When a hydrocarbon molecule is cracked, the number of hydrogen and carbon atoms remain the same but they have been rearranged. One part of the chain may now contain carbon atoms which are surrounded by *four* other atoms. No more atoms can be attached and it is said to be **saturated**.

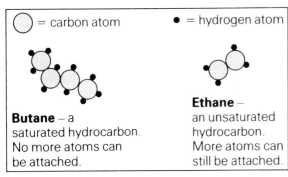

Butane – a saturated hydrocarbon. No more atoms can be attached.

Ethane – an unsaturated hydrocarbon. More atoms can still be attached.

Cracking produces saturated and unsaturated compounds.

The other part contains some carbon atoms surrounded by only *three* atoms. More atoms can still be attached to these carbon atoms, and it is said to be **unsaturated**. These **unsaturated** compounds are the useful products.

How are they useful?

Because more atoms can still be attached, the small unsaturated hydrocarbons are more useful than the saturated hydrocarbons. These small molecules, called **monomers**, can be joined together to form plastics or **polymers**. For example, the monomer **ethene** can form the polymer **polyethene** (polythene). Polythene can be made with chain lengths of 1000 to 20 000 carbon atoms. This is done by altering the conditions of **polymerisation**. Other polymers can also be made using different monomers. These polymers are often called **plastics** because they can be easily shaped during manufacture.

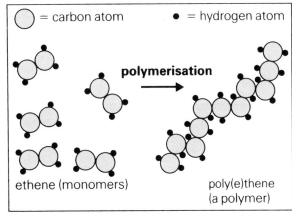

Different forms of plastics have different properties – each property gives rise to different uses.

What are plastics used for?

Today we really do live in a 'plastic' world...

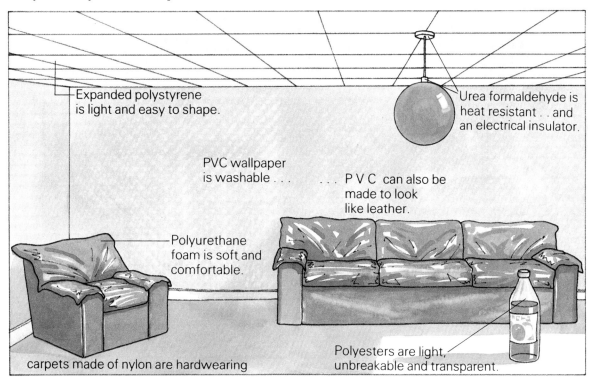

1. Look at the figure on the opposite page
   a. Which fractions are cracked after being distilled?
   b. Which fraction was produced in the greatest amount by cracking?

2. Why is heat needed for cracking?

3. What is the difference between a saturated and unsaturated hydrocarbon?

4. Draw the shape of the polymer formed by joining 10 ethene monomers. Explain why the polymer can still increase its chain length.

5. Plastic bottles for fizzy drinks are made of polyesters. What properties does it have which make it ideal for this use?

6. Why do some people think the use of crude oil as a fuel wastes resources?

5.13 Flammable materials

Burning up!

To burn a material you need *oxygen*, which is present in the air, and a source of *heat*. Materials, such as petrol or natural gas, which burn very easily are said to be **highly flammable**. Materials that don't burn are said to be **non-flammable**, such as most metals. Metals usually just melt when they are heated. Other non-flammable materials, such as limestone, break down into simpler substances.

*If materials are to act as **fuels**, they need **heat** and **oxygen**. Remove just one of these and the fire will go out.*

When materials burn . . .

When some materials burn they produce harmful substances. . .

Natural gas produces heat and harmless invisible gases such as carbon dioxide and water.

A lot of heat is produced from a burning sofa. In 3 minutes the temperature of the room can rise by 1000°C.

A lot of carbon in the form of soot or smoke is produced – make breathing very difficult.

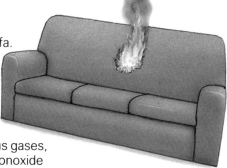

Invisible poisonous gases, such as carbon monoxide and hydrogen cyanide, are produced.

Soft padding used to be made from animal hair but up to now polyurethane foam was used instead. This foam may be cheap, but in a fire it is dangerous.

A controlled fire in the home is fine – an uncontrolled furniture fire can be lethal.

Nowadays, all new furniture sold in Britain has to be fire resistant. This should help to reduce the number of deaths in household fires.

Don't try to use water, **don't** try to take it outside like this. Cover with a damp towel. Remove the pan from the cooker. **Don't** remove the towel until the pan cools.

Chips are dangerous!

Many people have chips at least once a week, yet frying chips is the main cause of fire in the home. Fat or cooking oil are not very flammable substances, but when overheated they give off a blue smoke and can catch fire spontaneously. Alternatively, the oil or fat spills or spits over onto the cooker and catches fire.

A flammable test

Wool, as a carpet material, is now being replaced by artificial fibres such as nylon. One method of comparing how easily they burn is to place a hot metal nut on a carpet for 30 seconds and examine the size of the burn. Look at the results in the picture. What do they show you about the flammability of nylon and wool?

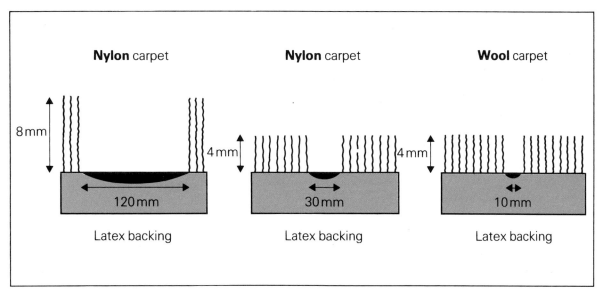

Predict how large the burn mark will be for a 'long pile' wool carpet.

Beating the flames

Some materials, such as asbestos, are very flame resistant and do not catch fire easily. Most materials, however, have to be treated with a flame retardant or flame proofing solution such as **borax** to make them safer. Burns, however, are not only caused by flames but by the heat produced by charred material still glowing when the flame has gone out.

Look at the table. What does it show you about the effect of washing flame proof materials?

	Untreated Cotton	Cotton treated with borax	Cotton treated with borax and washed
Flame test	Flame spreads immediately	Flame takes 10 seconds to spread	Flame takes 6 seconds to spread
Smoulder test	Charred material glows red for 30 seconds	Charred material glows red for 5 seconds	Charred material glows red for 12 seconds

Flame retardants, such as borax, make materials less flammable.

1. Look at the sign of the fire triangle. Why are the sides labelled heat, oxygen and fuel?

2. Why do petrol stations have 'NO SMOKING' signs?

3. Look at the diagram of the burning sofa.
   a Why are people often not able to escape from these fires?
   b What do you think are the most likely causes of these fires?

4. Look at the carpet burns shown above. What effect does a the pile height b the type of fibre have on the size of the burn.

5. Why do fire fighters sometimes wear asbestos suits?

6. Why do flameproof materials need to be retreated after washing a few times?

5.14 Materials – 'tailor-made' for

A car is made of many different materials. Each material has particular properties which make it right for the job...

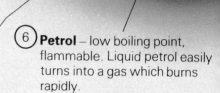

④ **Copper** – electrical conductor. This exce[llent] conductor makes sure no electrical energ[y is] wasted.

③ **Lead** – electrolysis in battery. A lead-acid battery produces electricity.

⑭ **Paint** – tough, attractive. Paint protects the steel from rusting.

① **Glass** – transparent. It lets the light out from the headlamp, but keeps water and dust out.

② **Plastic** (PVC) – electrical insulator. This makes sure the flow of electricity stays inside the wire.

⑮ **Chrome plating** – tough, attractive, non-rusting. Chrome protects the steel in places where paint would probably be chipped.

⑥ **Petrol** – low boiling point, flammable. Liquid petrol easily turns into a gas which burns rapidly.

⑤ **Aluminium** – lightweight heat conductor. An aluminium engine weighs less than a normal steel engine. It also conducts heat away more effectively.

your use!

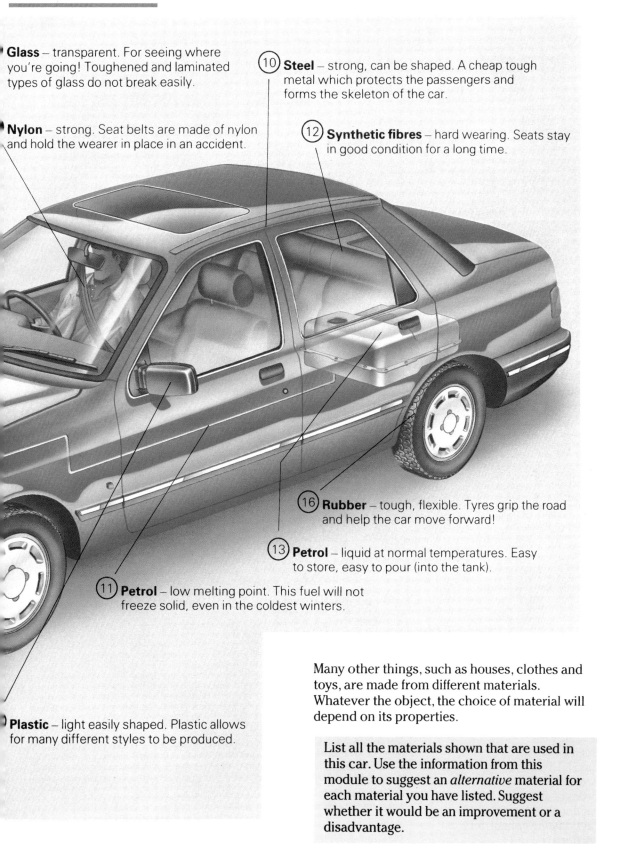

Glass – transparent. For seeing where you're going! Toughened and laminated types of glass do not break easily.

Nylon – strong. Seat belts are made of nylon and hold the wearer in place in an accident.

(10) **Steel** – strong, can be shaped. A cheap tough metal which protects the passengers and forms the skeleton of the car.

(12) **Synthetic fibres** – hard wearing. Seats stay in good condition for a long time.

(16) **Rubber** – tough, flexible. Tyres grip the road and help the car move forward!

(13) **Petrol** – liquid at normal temperatures. Easy to store, easy to pour (into the tank).

(11) **Petrol** – low melting point. This fuel will not freeze solid, even in the coldest winters.

Plastic – light easily shaped. Plastic allows for many different styles to be produced.

Many other things, such as houses, clothes and toys, are made from different materials. Whatever the object, the choice of material will depend on its properties.

List all the materials shown that are used in this car. Use the information from this module to suggest an *alternative* material for each material you have listed. Suggest whether it would be an improvement or a disadvantage.

MODULE 5 MATERIALS

Index
(refers to spread numbers)

A
abrasives 5.3
air (as insulator) 5.4
aluminium 5.1
antifreeze 5.1
atoms 5.3

B
bending 5.6
bitumen 5.11
boiling point 5.10
bone 5.1
borax 5.13
butane 5.13

C
casting 5.2
chrome 5.14
compost 5.9
condensing 5.10
conductors 5.4
copper 5.4, 5.14
crystal 5.3

D
damp, decomposers 5.9
density 5.5
diamond 5.3
diesel 5.11
ductile 5.2

E
elastic 5.2
electroplating 5.3
ethene 5.12
expanded polystyrene 5.5
extrusion 5.1
extension 5.1

F
fire 5.13
flame retardents 5.13
flame test 5.13
flexibility 5.6
flint 5.1
foam 5.13
forces 5.6
forces of attraction 5.8
fractional distillation 5.10
fractionating column 5.11
fractions (oil) 5.11

G
gas (state) 5.10
glass 5.14
   fibre 5.4

H
hardness 5.8
hardwearing 5.8
heat 5.4, 5.10, 5.13
hydrocarbons 5.11, 5.12
hydrogen 5.10

I
insulators 5.4

L
latex 5.13
liquid (state) 5.10
lubricating oil 5.11

M
melting point 5.10
mixed materials 5.8
molecules 5.9, 5.11
monomers 5.12

N
nickel 5.8
nylon 5.6, 5.13, 5.14

O
oil (crude) 5.11
oxygen 5.13

P
paint 5.1, 5.14
paraffin 5.11
petrol 5.1, 5.11, 5.14
plastic (property) 5.2
plastics 5.6, 5.7, 5.11, 5.12, 5.14
polyesters 5.12
polymers 5.12
pressing (of metals) 5.2

R
recycling 5.9
rolling of metals 5.2
rot (wet/dry) 5.9

S
saturated compounds 5.12
shape of materials 5.2
silver 5.4
smoulder test 5.13
solid (state) 5.10
states of matter 5.10
steel 5.1
stone 5.1
strength 5.6
   compressive 5.6
   tensile 5.6
stretching 5.2
synthetic materials 5.1

T
tension 56
thickness 5.2
tin 5.8

U
unsaturated compounds 5.12
urea fomaldehyde 5.12

W
wood 5.1, 5.8

Z
zinc 5.3

Photo Acknowledgements
The references indicate spread numbers and, where appropriate, the photo sequence.

Barnaby's *5.4/2*, (Micky White) *5.5/1*, (Norman Price) *5.5/2*, (John Edwards) *5.6/2, 5.11/1*; British Telecommunications *5.11/1*; J Allan Cash *5.14*; John Cleare Mountain Camera *5.6/1*; Fire Research Station *5.13*; Sally & Richard Greenhill – contents *5.2(×2), 5.3/2*; GeoScience Features *5.3/3, 5.9/1, 5.9/3, 5.9/4*; Adam Hart-Davis *5.10/3*, Trevor J Hill *5.5.5/3, 5.8/2, 5.10/3, 5.11/4*; Frank Lane Agency (Mark Newman) *5.4/1*, (Walter Rohdick) *5.9/2*, (Mike Thomas) *5.10/1*;

Picture research: Jennifer Johnson

MODULAR SCIENCE for GCSE

MODULE *Metals*

Metals *are a particular group of materials. They all have similar physical properties– a shiny appearance, good conductors of heat and electricity, usually strong and flexible. This module looks at the physical and chemical properties of metals, and the many uses which we have for them.*

Relevant National Curriculum Attainment Targets: (6), (7), (8), (10)

6.1	Metals are useful . . .
6.2	The metal for the job
6.3	Making new metals
6.4	Improving iron
6.5	How do metals react?
6.6	Metals corrode
6.7	Titanium – the supermetal
6.8	Metals from their ores
6.9	Case study: iron
6.10	Case study: aluminium
6.11	The rarer metals
6.12	Metals and the environment
6.13	Spaceship Earth
6.14	Will we still need metals?
	Index/Acknowledgements

MODULE 6

6.1 Metals are useful...

...in the home

...for carrying electricity

...for looking good

Scissor blades are made from hardened steel. *Why not make scissors from gold or silver for the very rich?*

Light bulbs have a tungsten filament. *That's rather pricey, isn't it? What's wrong with the nichrome wire that's used in electric fires?*

Mirrors have aluminium sprayed onto the back of the glass. *Why isn't silver used any more?*

Gold, silver and platinum are used for jewellery. *What's wrong with brass or tin? They look shiny enough on the market stalls!*

...and for all sorts of useful things

Most of its bodywork is made from alloys of aluminium. *That's the same as cooking foil! Why not use steel? Isn't it much stronger?*

Concorde has a titanium nosecone. *That's expensive, isn't it?*

Titanium is also used (in the form of its oxide) as a white pigment in paint. *What happened to the 'white lead' paint that used to be around?*

Cutting tools and drill bits are made from tungsten steels. *What's so special about these types of steel?*

You can find out the answers to all these questions – and many more uses for metals by finding out all about metals.

6.2 The metal for the job

So why are different metals chosen for different uses? The answer lies in their wide range of chemical and physical properties.

Melting points

Mercury is used in thermometers because it is the only metal that is a liquid at room temperature (about 25°C). At the other end of the scale, **tungsten** is used for light bulb filaments because it can get 'white hot' without melting.

Density

Some metals such as **lead** are very heavy for their size – **dense**. This means that they are useful for holding things down (as weights). They would not, however, be much good for aeroplanes – for these you need 'lightweight' metals such as **aluminium**.

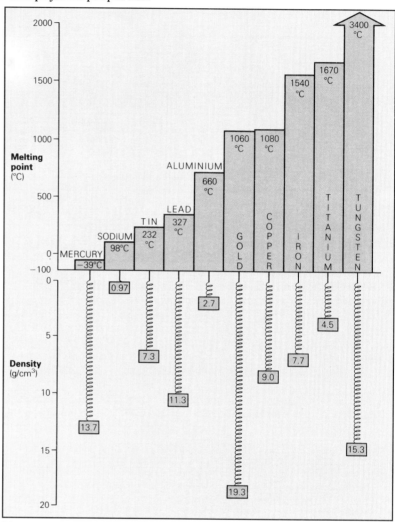

Melting point and density influence the use of many metals. Why can't you use sodium in an ordinary thermometer?

Bridges are supposed to stay in one place – so they are made from cheap strong steel instead of expensive titanium alloys.

Strength

Many of the common uses of metals, in construction or for machinery, rely on the fact that metals can be very strong. **Iron** and the various types of **steel** are important because of this. Just think of how much iron and steel there is in use around the world – in cars, trains, ships, bridges, railway lines ... the list is endless. Iron is also used to reinforce concrete in most modern buildings.

Other metals are also used for their strength, but usually in the form of carefully blended mixtures – **alloys. Aluminium**, **magnesium** and **titanium** are widely used for aircraft because they combine strength with low density – essential if you wish to 'get off the ground'!

Heat and electrical conduction

Aluminium and **copper** are good conductors of heat, which is why they are used for saucepans – among other things. They are also good conductors of electricity, which is why they are used for electrical wiring.

Squash and stretch

Metals can be squeezed through giant rollers to make plating, or pulled out through holes to make wire. **Gold** is the best of all at this. It can be beaten into 'gold leaf' – so thin that you can even see through it – or drawn into wires as thin as a spider's web.

These four rollers can squeeze large blocks of metal into long thin wires.

All that glistens...

Gold is also a very **unreactive** metal, and so does not tarnish in the air. Because of this, it keeps its shine and so is highly prized for jewellery.

Most metals do tarnish, however, because they react with the oxygen in the air. The dull oxide layer on **aluminium** actually protects the metal beneath, but with **iron** and **steel** it eats in and destroys the metal as rust.

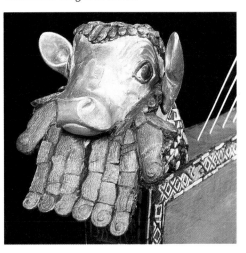

This figure was made thousands of years ago – but gold remains shiny, long after other metals have tarnished.

Metal	Grammes of metal found in 1 tonne of earth	Price per tonne of metal
Aluminium	70 000	£750
Iron	40 000	£130
Zinc	70	£500
Copper	45	£1000
Lead	15	£290
Tin	2	£9000
Silver	0.07	£150 000
Gold	0.004	£8 000 000

...costs a lot!

Silver is a better conductor than **copper**, so why don't we use it for our wiring at home? **Gold** is even denser than lead but is not poisonous, so why don't we use it for fishing weights? The answer to this and many other similar questions is simple – cost! When deciding on the metal to use for a particular job the properties are obviously very important, but the choice must also be 'cost-effective'.

1 List the metals shown in order of increasing melting point.

2 What are the densities of gold, lead, copper, iron and aluminium?

3 Look at the table showing the price of metals. How does the price relate to the ease of finding the metal? Which **two** metals do not fit the pattern?

4 Look for metals in use around you, and try to work out why they are used.

6.3 Making new metals

Bronze is beautiful!

When copper was discovered about 7000 years ago, it was used to make pots, and jewellery, but it was too soft to make knives or spears. Tin had also been far too weak a metal to be of any use. Yet when the two were accidently mixed, the resulting metal was to change the world.

The new metal, a mixture of copper and tin, was called **bronze**. It could be made into knives which could be sharpened, and into pots that were tough and hard-wearing. The Bronze Age had begun!

Mixed-up metals

By mixing different metals in different proportions, new metals may be made. These mixed metals are called **alloys** and have a wide range of properties. As was the case with bronze, these new metals are often more useful than the pure metals themselves. The discovery of an important new alloy has often helped us to develop our technology further. It is now possible to 'design' a suitable alloy for most jobs, but this is not a simple process. Sometimes it is even necessary to add other elements which are not metals.

Bronze was a much sought-after alloy – it could be used for tools, pots and even statues.

Hot stuff – alloys and melting points

Often an alloy behaves quite differently from the metals from which it is made:

Solder Lead melts at 327°C, tin at 232°C. You might expect that if you mixed these two, the alloy would melt somewhere between the two figures. In fact, such an alloy can melt as low as 183°C, which is useful for joining wires in electrical circuits together. A typical **solder**, has about 20% tin.

Sometimes the difference in the melting point of the metals poses special problems when making an alloy:

Brass This is a common alloy of copper (70%) and zinc that is used for its hardness and appearance. The problem in making it is that copper melts at 1083°C, while zinc boils at only 907°C. If they were just melted together, the zinc would boil off and burn up! This is overcome by melting the copper first, and then dropping the zinc in quickly as tiny pieces, but some zinc is still lost.

Making brass has its difficulties, but the result is very attractive.

Tough stuff – alloys and corrosion

Some alloys have been made to resist corrosion:

Gun metal Copper and tin make **bronze**, copper and zinc make **brass**. Combining all *three* (copper, tin and zinc) make **gun metal** – a tough, hard-wearing metal that resists corrosion and is used for ship fittings, as well as guns!

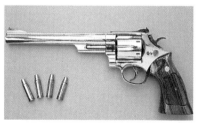

Guns need to be made from a special alloy, gun-metal. Bullets need different properties and so they have to be made of other alloys

Cupro-nickel Copper and nickel form a resistant alloy (cupro-nickel) that can cope with being passed through many sweaty hands without tarnishing, and so is ideal for coinage! The addtion of zinc to a copper and nickel mix makes '**German silver**', which is used for ornaments and car trims.

Strong stuff!

The old saying about 'the weakest link in the chain' does not hold for alloys. Mixing 'weak' metals together can have surprising results!

Duralumin Aluminium has a very low density for a metal, but is too weak to be of much use in construction when pure. Magnesium is even weaker, yet a combination of the two produces a strong alloy. The addition of a small amount of copper improves things even further, making **duralumin**, which is used in aircraft construction.

Other alloys

Type metal Antimony is an unusual metal in that it expands slightly when it solidifies (as water does when it freezes). When mixed with lead, it makes an alloy which can be cast in a mould to produce sharp, clear printing, typewriter keys which can then press firmly against the typewriter ribbon.

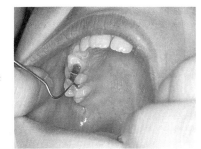

Old typewriters needed striking 'letters' made of type-metal if they were to give sharp and clear printed letters.

Amalgam When silver or tin are mixed with liquid mercury, they make a putty-like alloy called an **amalgam**, which hardens if left for a while. Great for filling holes in teeth!

What next?

The numbers of possible alloys are endless, and research is constantly in progress to find new forms. Some combinations have as yet proved impossible to make, however, due to conditions on Earth. The difference in the densities of lead and aluminium is so great, for example, that they separate out like oil on water when melted together. For this reason work on new alloys has been undertaken during Skylab missions to take advantage of the 'weightless' conditions in space. Perhaps one day there will be orbiting factories producing new alloys for use back on Earth!

'Open wide – I just need to squeeze some amalgam into the hole in your tooth'

What about iron?

The alloys of iron are so important they have a section of their own...

1 What metals are mixed to make bronze? Why was bronze so useful?

2 What's so odd about the melting point of a lead/tin alloy? What is this alloy called? What is its use?

3 Why is brass difficult to make? How is it done?

4 What properties of cupro-nickel make it suitable for coins? What metal is added to make 'German silver'?

5 Why are aluminium alloys so suitable for aircraft construction? What other metals are added?

6 Explain why lead-aluminium alloys might be made in space, but not on Earth. Can you suggest what the properties of these alloys might be?

6.4 Improving iron

Even a concrete bridge needs lots of steel – to reinforce the pillars and to support the bridge during building!

The steel age!

Pure iron is not particularly strong, not particularly hard, and rusts easily – no big deal really! Steel, however, can be hard, strong, or even 'stainless'. Without steel, we would have no cars or cans, trains or tower blocks. More than 90% of all the metal used is steel.

What is steel?

Just iron alloyed with a few percent (or less) of other things! The odd thing is that the chief ingredient is not a metal at all, but carbon – the same stuff that makes up brittle charcoal!

How much carbon?

The simple answer is 'not a lot'. The graphs show how the strength and hardness of steel changes with the carbon content. Between 0 and 1% of carbon, the strength trebles; the hardness also increases. **'Mild' steel** is bendable in thin sheets and is used for car bodies, 'tin' cans for food and ships' hulls. **'Medium' steel** is harder, and is strong enough for use in hammers or axe heads.

However, above 1%, the *strength drops* a little, while the *hardness still increases.* Scissors, knives, chisels and files are made from this **'high carbon'** steel because the harder the steel, the sharper the cutting edge.

Alloys with a higher carbon content are not usually produced 'on purpose', but iron straight from the blast furnace has a high carbon content from the coke used in the process. This can be poured straight into a mould to make hard but brittle **cast iron**.

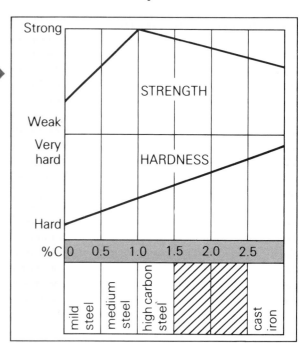

Cast iron was used to build the Ironbridge in Shropshire – as well as millions of manhole covers!

Adding other metals

Adding other metals can further change the properties of steel. At the turn of the century, it was discovered that adding a few percent of nickel produced a tough and resistant alloy ideal for cogs and **gears** – as well as armour-plating!

Adding up to 20% of tungsten to steel also makes a very hard alloy, but this one stays hard even when hot. This alloy is used for high speed **drills**, lathe tools and files.

Adding chromium (13–20%) to steel has a different effect. This makes the rust-resistant **'stainless** steel' used in cutlery, razor blades and power station steam piping – the best (but most expensive) way to beat corrosion!

Wrought iron

The 'raw' iron from a blast furnace contains carbon and other impurities which make it hard but brittle. Nowadays this is 'burnt out' with oxygen in high-temperature furnaces, but in the last century these processes had not been perfected. The early furnaces produced only spongey masses of iron with all the impurities still mixed in – slag. This had to be beaten with hammers to forge the hot metal into shape. The resulting metal was called **wrought iron**. It still contained some impurities but was very strong. It was used for ornamental ironwork, chains, and even the Eiffel Tower!

Making wrought iron needed great skill and much hard labour. The 'wrought iron' gates and railings made today are usually only twisted mild steel. *Real* wrought iron can be most easily spotted when it rusts – the layers of slag left in it are exposed, making the surface look like puff pastry'

Heat treatment

If a metal is examined under the microscope, it can be seen to be made up of tightly interlocking crystals. The properties of the metal depend on this crystal structure.

The structure of steel changes if it is heated, and this can be used to change the properties. If a piece of steel is heated until it is red hot and then cooled rapidly – **quenched** – by plunging it into cold oil or water, it becomes hard and brittle.

If this is then reheated to red heat and then allowed to cool slowly, it becomes soft and bendable again. This is called **annealing**.

If it is now cleaned and heated gently until it turns a light brown colour, and then quenched, it will become tough and springy – **tempered.**

Zinc under a microscope. You can see the patterns of the crystals of zinc.

1 What is added to iron to make steel?

2 How much carbon is there in mild steel? What is it used for?

3 What type of steel would be used for a file? How much carbon might it contain? Describe this type of steel.

4 Why does cutlery steel contain chromium?

5 To improve the properties of each tool, what heat treatment might have been given: **a** to a file; **b** to a needle?

6 Jane sterilised a needle over the gas before using it to remove a splinter. When she later tried to sew with it, it snapped in half. Explain what happened.

6.5 How do metals react?

Metal	Percentage of crust	First discovered
Aluminium	7	150 years ago
Iron	4	3000 years ago
Sodium	2.5	80 years ago
Magnesium	2	120 years ago
Zinc	0.007	2000 years ago
Copper	0.0045	7000 years ago
Lead	0.0015	7000 years ago
Tin	0.0002	6000 years ago
Gold	0.0000005	10000 years ago

Discovering metals

Look at the table. From the composition of the Earth's crust alone, which metal would you have expected to have been discovered first? If you said iron or aluminium, it would make sense as they are the commonest metals. Yet if you study the table, you will see it was gold – which is very rare – that was discovered first. Why was that?

Powdered iron will burn – sparklers are a good example of this.

Getting a reaction . . .

Let's look at the way metals react with air, water and acids – this may help to explain the order of discovery.

. . . in air

Some metals, such as sodium, will catch fire spontaneously in air. Others, such as magnesium, may be easily lit in thin strips, and burn vigorously. Iron, zinc and aluminium are not usually thought of as things that will burn – but if finely powdered, they too will flare up. Copper dust will not burn, but it still does react with the oxygen in the air, turning black. Gold and the other 'noble' metals do not react at all.

. . . in water

Sodium reacts furiously with water, whizzing round on the surface and getting so hot that it melts to a silvery ball. Hydrogen gas is given off and, if the ball is 'trapped' on filter paper, this catches fire. The reactions of other metals are less spectacular. Magnesium will bubble with hydrogen in hot water, but the others react much more slowly or even not at all.

A bright yellow light is given off if sodium catches fire when it gets too hot reacting with water.

. . . in acids

Hydrogen gas is also produced when metals react with acids, but this reaction is much faster than the reaction with water – sodium would be too dangerous to try! Magnesium reacts very vigorously, iron and zinc more slowly. Lead and tin will only react with warm, strong acids, while copper and gold will not react at all.

Magnesium, zinc and copper all react differently with acids – they have a different **reactivity**.

Getting in order

So different metals react at different rates, and can be listed in an **order of reactivity**. This order is the same, no matter what the reaction and it ties in with the order of discovery of metals. When metals burn, they react with the oxygen in the air to form new **compounds** – the metal oxides. The more vigorous the reaction, the 'stronger' the compounds they make. Some metals are much more reactive than others – their compounds are more difficult to break apart. Aluminium and iron are common in the Earth's crust, but they are so reactive that they are normally tightly locked up in compounds. Copper and lead are also found in compounds, but these are easier to break down. Gold is so unreactive that it may be found 'native', as the pure metal. This is why gold was the first metal to be found – it was just waiting to be discovered!

Reactivity and discovery

So most metals are found in the ground but are locked up in compounds. These are the mineral **ores.** The metals that were discovered first were those that could be most easily released from their ores. Lead was discovered early, because the heat from an ordinary cooking fire was enough to free it from its ore. These first discoveries were 'accidental' – nobody knew what metals were 'hidden' in the compounds.

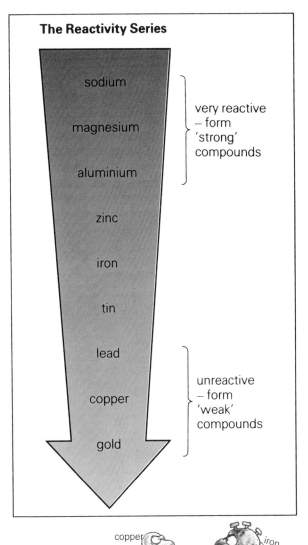

The Reactivity Series

sodium
magnesium
aluminium
— very reactive – form 'strong' compounds

zinc
iron
tin
lead
copper
gold
— unreactive – form 'weak' compounds

Getting the metals out

If a clean iron nail is dipped into copper sulphate solution, it comes out 'copper plated'. The iron is more reactive than the copper and so 'pushes' the copper out of its compound, taking its place in the compound – like a sort of 'chemical bully'! This idea is very useful, as *metals can always be 'pushed out' of their compounds by metals above them in the reactivity list.* Iron can be pushed out of its oxide by aluminium powder. The reaction needs a 'kick-start' of energy from a fuse to get it going but, once started, it gives off enough heat of its own to melt the iron produced. This is known as a 'thermit' reaction. It is spectacular, but dangerous if you get too close as it can spit out droplets of molten iron! Apart from livening up the lab this reaction was used to weld iron railway lines together.

A 'chemical bully' at work.

 1 Draw a separate 'order of reactivity' lists for the reactions of metals with air and acids. Compare them.

2 If a metal is very reactive, will its compounds be weak or strong?

3 Why was lead discovered so early?

4 Draw up a list of metals in 'order of discovery'. Compare this with your 'order of reactivity'. What do you notice?

6.6 Metals corrode

Painting the Forth Bridge

'It's like painting the Forth Bridge' is a phrase that's often used to describe a never-ending job. The Forth Bridge has to be painted to stop it from rotting away – **corroding** – but this is a very long job. In fact, the bridge is so big that as soon as the painting is finished at one end, it's time to start again with fresh paint at the other!

Iron rusts

The Forth Bridge is made from iron. As all car owners will tell you, iron – whether 'pure' or in the form of steel – rusts! This means that the iron is reacting with both the oxygen in the air *and* any water present to form a new, stable compound. This then swells and blisters away from the surface, exposing fresh iron, and so the reaction carries on. Eventually, the whole structure collapses into dust – all too quickly with a lot of cars!

Iron lamposts often corrode badly at the base and collapse due to a particular habit of dogs!

Rust blisters away, allowing more water and air to react with the iron.

Salt water speeds up rusting so cars kept at the seaside may rust faster than those inland. Salt used on icy roads in winter will have the same effect.

How do we stop rusting?

Both water and air are needed for rusting to occur. Paint seals off the metal surface and so stops rust. But if this protective coating is chipped or scratched, rust will soon start to work underneath it. Rusting will continue until all of the rust is removed and the metal is cleaned and repainted.

New cars have several layers of special paint beneath their 'top coat' to stop rusting. The underside is very open to rusting, so this is **undersealed** with a layer of bitumen. Coatings of oil or grease also keep out air and water and so prevent rusting.

Steel plating is very useful for making many things, but rusts very easily. It is often protected by coating it with a thin layer of another metal, such as tin (for tin cans) or zinc for things like barbed wire.

Paint dipping and tin plating – two ways to protect steel.

Reactivity again

If the tin coating of a food can is broken, the iron rusts very fast indeed – yet if a zinc coated sheet is scratched, it is the zinc that corrodes! This is because when two metals are joined by water they form a simple electric cell. This causes the more reactive metal to corrode, while leaving the less reactive one intact. Zinc is *more* reactive than iron, tin is *less* reactive than iron. So it is the iron in the tin can that rusts, but the zinc on the sheet that corrodes in the other case.

A noble sacrifice!

This difference in reactivity has been used to protect iron from rusting in a different way. Ships sometimes had blocks of zinc bolted to their hulls which then corroded in place of the iron of the hull. These 'sacrificial' blocks were much easier to replace than the hull itself!

If the designers of the QE2 had forgotten this, they would have found that its reactive aluminium decks would 'protect' its steel hull in the same way! (Fortunately they remembered and fitted a plastic layer between the two, so it's quite safe!)

Why doesn't zinc corrode?

Zinc is more reactive than iron, so why doesn't it just corrode on its own? In fact it does react to form an oxide layer. But unlike iron this does not expand and blister off. Instead it forms an unreactive outer layer that protects the metal beneath. The same thing happens with highly reactive aluminium, which is why kitchen foil does not corrode.

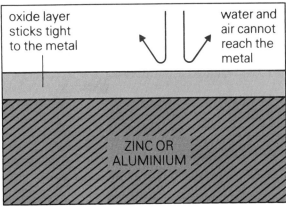

'Self-sealing' aluminium and zinc protect themselves with a layer of oxide on their surface.

1. What would happen to the Forth Bridge if it were not painted?

2. Why doesn't aluminium corrode? Why doesn't rust protect iron in the same way?

3. Some cars have aluminium engines. What would happen if these were bolted directly onto the steel body of the car?

4. How are cars protected against rust? Can you think of any reasons why manufacturers do not make their cars of totally rust-proof stainless steel?

5. Why is it not good enough just to 'paint over the rust' on an old car?

6.7 Titanium – the supermetal

It's tough . . .

So what is so super about titanium? Well, it has certain properties which make it an excellent metal for supersonic aircraft construction. Firstly it is as strong as steel, yet has only half the density – which means that the same sized piece weighs only half as much! Secondly, it has a very high melting point (nearly 2000°C) and so can cope with the effects of frictional heating caused when flying at high speed.

The new American spy plane Blackbird is designed to fly at supersonic speeds – 3000 km per hour. Ordinary aircraft materials would tear apart at such speeds, so it has been built almost entirely of titanium.

Thirdly, it is highly reactive, and combines instantly with the oxygen in the air. This very reactivity makes pure titanium almost totally resistant to rust or corrosion. The instant it comes into contact with air it reacts to form a thin oxide layer, which prevents further damage – a tougher version of the layer that forms on aluminium and zinc.

. . . it's long lasting

This 'rust-proof' property has led to titanium finding important uses in many different areas. In nuclear power stations, for example, it is used for the pipes which carry steam from the reactors to drive the generators, and in the condensers that turn the steam back to water for recycling. It is also used for piping in conventional power stations where sea water is used as a coolant, which would eventually rust even 'stainless' steel.

The human body is a 'high-risk' environment, too! Many arthritis sufferers have benefited from replacement hip joints (shown here held by a surgeon), in which the artificial 'ball' of the joint is made from titanium as it is very strong but, just as importantly, will not be attacked by corrosive body fluids!

Finding, mining . . .

Titanium is a relatively common metal (0.6% of the crust) but, like aluminium, is not found in commercial quantities in veins. Its oxide ore is hard and resistant and so is not broken down by weathering or erosion, and collects, like quartz, in sand. Titanium bearing sands are found in India, Japan, Malaysia and Central America, but the most important deposits of all are the beach sands of Australia. Yet even here, the ore amounts to only 1% of the sand, so large scale separation methods are needed.

This whole process has to be carefully controlled to minimise environmental damage. The topsoil is first removed and then later replaced over the worked sand.

The huge processing plant, is built to float on an artificial pond created by digging a huge hole in the sand.

Up to 1500 tonnes of sand is cut or sucked from one end of the pond every hour.

The ore is separated from the sand on the platform.

ore concentrate pumped ashore

The unwanted sand is then used to 'backfill' the pond, and the whole operation moves steadily along the beach.

. . . and refining titanium

Titanium is such a very reactive metal that it would immediately recombine with oxygen if it were produced in air. This makes its production difficult and expensive, as the whole process has to take place in sealed, air-proof units in an atmosphere of totally unreactive gas such as helium or argon. Yet the basic reaction used to free it is a simple use of the 'chemical bully' idea.

The titanium oxide ore is first converted to titanium tetrachloride (this lets the main reaction run at much lower temperatures), which is then allowed to react with sodium. The more reactive metal pushes the titanium from its compound in a violent thermit reaction, to form titanium and sodium chloride.

Even then, the problems of refining are not over! The metal is in a spongy form, and has to be melted by heating to 2000°C in an electric arc furnace before it is ready for use.

---

1 Why was titanium used for the 'Blackbird', but is not used for all aircraft?

2 What property makes it useful for nuclear pipes and replacement hip joints?

3 How is the environment protected when titanium is mined?

4 Describe the refining process for titanium. Why is it turned to the chloride first?

6.8 Metals from their ores

Spoilt for choice?

Metals can be extracted from their ores in different ways. Thermit reactions work well, but are not usually used to extract metals commercially. Can you think why not?

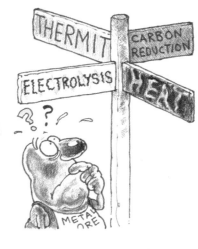

What's the problem?

Let's look at the case of one of the earliest metals to be discovered – lead. It's easy to get the unreactive lead from **galena**, it's ore. You just have to mix it with a more reactive metal such as aluminium and it will be 'pushed out' in a violent thermit reaction ... but aluminium, being reactive, will be much harder to get from its compounds than lead, so that means using a 'hard earned' metal to get an unreactive metal like lead.

Fortunately non-metals fit into the reactivity list as well. Carbon fits in between iron and zinc and so can be used to push lead from its compounds by a **reduction reaction**. Unlike aluminium, however, carbon is in good supply in a (more or less) pure form as charcoal (or coal). In the example of lead, then, the galena is roasted in air to make lead oxide and then reduced with carbon:

lead oxide + carbon → lead + carbon dioxide

Early discoveries

As lead has a low melting point, it could have easily been produced 'by accident' in the charcoal-rich embers of a cooking fire made on ground rich in galena. Copper could also have been discovered in this way, but an ordinary working fire would not have been hot enough to melt it. Its discovery 7000 years ago was probably due to the use of pottery kilns which got much hotter than ordinary fires – hot enough for carbon to release copper from its ore.

Once copper was discovered, due to the improved kilns of the Egyptians and Mesopotamians, many uses were soon found for it.

Next came iron

Iron can also be pushed out by carbon, but much higher temperatures are needed. Lead melts at 327°C, copper at 1083°C, but iron needs a temperature of 1535°C! An ordinary fire will not reach anywhere near this temperature unless air is blown through it, to increase the supply of oxygen. So, despite the fact that iron ore is common, iron was only discovered 3000 years ago. Though not very pure, iron was much better than bronze – so the Iron Age began! For many centuries, furnaces were still unable to fully melt the iron, which came out in the form of a spongy, impure mass. This then had to be beaten (**wrought**) into shape. Nowadays, iron is produced (in a fully molten form) in **blast furnaces**.

Though badly rusted, the iron blade of this Viking sword proves that medieval fires must have been hot enough to melt iron.

A different method is needed

This type of reduction reaction – using carbon to push the metal from the ore, is a good and 'cost-effective' way to extract many metals. But it doesn't work for metals like aluminium which are more reactive than carbon. Aluminium compounds have to be separated using large amounts of electrical energy, in a process called **electrolysis**.

Using electricity

One of the simplest examples of electrolysis, often seen in school laboratories, is that of copper sulphate solution. If two wires connected to a battery are dipped into this blue solution, the negative wire (the **cathode**) gets coated with fresh, metallic copper. This is because the copper in solution has a positive charge and is pulled towards the negative cathode (unlike charges attract). Here the charges cancel out, and metallic copper is deposited.

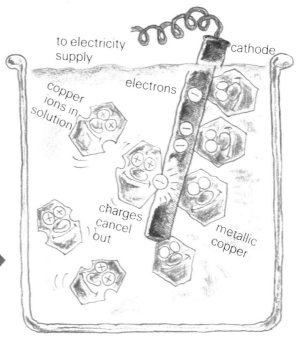

Electrolysis – using electricity to separate metals from their compounds.

Strong but light – aluminium makes good tennis rackets.

It's not easy with aluminium

Aluminium is extracted by the electrolysis of its molten ore, as it is too reactive to get from solution. Unfortunately, the common aluminium ore, **bauxite**, has a very high melting point, so vast amounts of energy are needed, first to melt and then to split the ore. This makes aluminium production difficult and expensive (see 6.10). This most common of metals was only discovered in the 19th century, and was extremely expensive to make; Napoleon's best dinner service was aluminium!

Which method is the 'best'?

Electrolysis could be used to get most metals from their ores. On the face of it, it might seem to be the simplest and most straightforward method of refining. But, as has been said, a lot of energy is needed, and this all adds to the cost of producing the metal. Energy is sold in 'units', each one being enough to keep a one-bar electric fire burning for one hour (1 kilowatt-hour, kWh). The table shows how many units of energy are needed to make 1 tonne of aluminium, iron and lead. Now you can see why carbon reduction methods are used whenever possible!

Metal	Energy needed to produce 1 tonne of metal
Aluminium	14 000 kWh (electrolysis)
Iron	350 kWh (carbon reduction)
Lead	100 kWh (carbon reduction)

 Why isn't aluminium used to make lead commercially?

 Why isn't electroysis used to get iron and copper from their ores?

2 Carbon will push iron out of its ores, so why was iron discovered so much later than lead?

6.9 Case study: iron

A very useful metal

As the table of production figures shows, iron is very definitely 'top of the league' of useful metals! Its production in the blast furnace through carbon reduction, is worth a closer look . . .

What is needed?

Iron ore Iron is common (4% of the crust). Its ores are found almost everywhere. The commonest ores are the oxides, hematite and limonite, which may be found in Britain, but the biggest and purest deposits are found in ancient 'banded iron' sediments in Canada, South America, Africa and Australia.

Coke Carbon is needed both to fire the furnace and to reduce the ore. At first charcoal was used, but modern furnaces use coke – coal which has been heated to drive off the 'volatile' oils and gases leaving fairly pure carbon.

Limestone Most iron ore contains a large amount of sandy material, silica, so limestone, a common rock made of calcium carbonate, is added to help purify the iron.

Metal	Production (million tonnes /year)
Iron	400
Aluminium	12.5
Copper	8
Zinc	6
Chromium	6
Lead	3
Nickel	0.7
Tin	0.2
Silver	0.01
Gold	0.001

Siting a blast furnace, then . . .

Up until the 18th century, iron making was a small-scale operation using charcoal, which was produced from wood by carefully controlled burning. For some time, Ashdown Forest in South-east England was an important centre, using the local ironstone. As a result of this, Ashdown Forest today has a distinct lack of trees!

In 1709 Abraham Darby discovered how to make iron using coke instead of charcoal, and this opened the way to iron-making on a much larger scale. Iron was needed for the new technology of the Industrial Revolution. The need for large supplies shifted the industry to the Forest of Dean, South Wales and the Welsh Borderland, where all three raw materials were to be found close together. Transport costs were as important then as they are today!

. . . and now

In the boom years of the 1960s, iron production reached a peak of nearly 600 million tonnes a year, but has now dropped back to 400 million tonnes. There is now something of a glut of iron and, in this 'buyers market', prices have fallen. Only the largest mines with the best ore have been able to keep going. It is now cheaper to import giant shiploads of iron ore from Australia than it is to use British iron ore! Now the main factor influencing the site is the nearness of a harbour.

Iron ore is transported in ships like this – nearly big enough for a game of football!

The blast furnace in action

In simple terms, the hot reactive carbon in the blast furnace is reducing the iron oxide – pulling the oxygen away from the iron. In reality, this happens in three stages.

First stage The coke burns, giving off lots of heat.
carbon + oxygen → carbon dioxide + heat energy
Second stage The carbon dioxide reacts with more carbon.
carbon dioxide + carbon → carbon monoxide
Third stage The carbon monoxide reduces the iron oxide.
carbon monoxide + iron oxide → carbon dioxide + iron

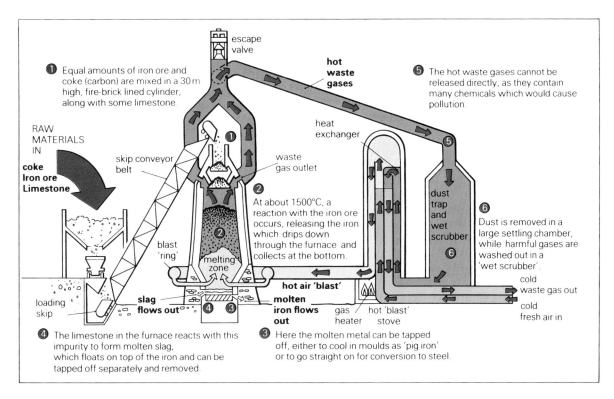

Waste not . . .

The hot air blast is produced by heating air in a sort of gas stove. But once the furnace is running, great energy savings can be made by preheating the fresh air in a heat exchange unit, using the hot waste gases and so recycling a lot of the heat energy.

1 What are the three raw materials needed to run a blast furnace?

2 Why was South Wales such a good area for iron production after 1709?

3 Why is limestone added to the ore/coke mixture?

4 What reaction keeps the blast furnace at 1500°C?

5 How are the waste gases treated and used?

6 Why is it necessary to have an escape valve?

7 Why do you think blast furnaces usually work continuously all year around?

6.10 Case study: aluminium

It's all too much!

Aluminium, like iron, is a common metal (7% of the crust) that is produced in bulk (12.5 million tonnes per year). But there the similarity ends. Aluminium is too reactive to be produced by carbon reduction, and so must be produced by the 'energy guzzling' process of electrolysis. Therefore aluminium, though cheap as an ore, is expensive to produce.

What is needed?

Aluminium may be common, but it does not form concentrated ore veins. Instead it is spread evenly throughout the soil tightly locked up in clay. There is probably tons of aluminium in the mud in you local area – but just you try to extract it! Fortunately, under tropical weathering conditions, clay breaks down to form **bauxite** – aluminium oxide.

Bauxite is fairly pure aluminium oxide, a 'high grade' ore that produces 50% metallic aluminium by weight. The problem is that it needs to be molten for electrolysis to work, and its melting point is over 2000°C. Working at such high temperatures could double the costs but, fortunately, by adding a little **cryolite** (a less common aluminium ore), the melting point can be lowered to 950°C.

Aluminium is a useful metal. It is light and strong, but it is expensive to produce. It is processed commercially, on a large scale, through high temperature electrolysis. . . . so a lot of energy is used.

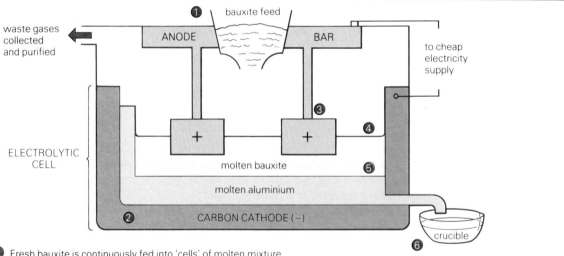

① Fresh bauxite is continuously fed into 'cells' of molten mixture, and molten aluminium is produced at the rate of 1 tonne per day.

② The lower part of the cell is lined with carbon that is connected to the electrical supply to form the cathode (−).

③ The anode (+) is made of carbon blocks which are lowered into the molten mixture. As the electric current flows (at nearly 150 000 amps per cell!) the aluminium oxide is torn apart.

④ The negatively charged oxygen is drawn to the anode where it reacts with the carbon, forming carbon dioxide. Because of this, the carbon blocks must be constantly replaced.

⑤ The postively charged aluminium ions are drawn to the cathode, where they turn to molten aluminium.

⑥ The molten aluminium is siphoned off into great crucibles, capable of holding 5 tonnes at a time. Once it has cooled to near its melting point (660°C), it is poured into moulds to make 50 kg 'pigs' or 500 kg 'sows'.

At what cost?

Because of the high production costs involved, aluminium plants tend to be very large – it is cheaper (per tonne) to make a lot rather than a little. A typical plant may have 300 cells like the one shown opposite. The plant at Fort William in Scotland uses 200 megawatts of power – equal to the output of a typical power station. Indeed, most plants have a power station of their own!

Siting the plant

This 'high cost' factor also means that the siting of the plant is crucial, as even minor additional costs could make all the difference to commercial success. The critical factors are:

Transport costs The plant must have easy access to a sea terminal large enough to take the bulk carriers that ship the bauxite in from abroad. It must also have suitable rail or road outlets for its refined metal.

Power costs A very large, continuous and reliable power supply is needed which is as cheap as possible! This means siting the plant near mountains (hydroelectric power), coal mines, oil or gas fields or a nuclear power station.

A quick glance at the tables will show why most plants are built near hydroelectric supplies – for good financial reasons!

Can you see the pipes running down the far hill? These form part of the hydroelectric power supply for this aluminium factory.

Fuel	Cost per kWh (p)	% of aluminium plants using the fuel
Coal	2.3	21
Oil	2.5	11
Gas	2.2	10
Nuclear	1.8	3
Hydro-electric	1.2	52

The supply of bauxite

The major deposits of bauxite (except those in Australia) are found in the developing countries. This might seem to be a good thing for such countries, but it can lead to problems. Jamaica, for instance, has two main ways of earning foreign currency – bauxite and tourism. They need to export as much bauxite as possible, but that means digging ugly holes in the ground and piling up spoil heaps – hardly a tourist attraction! Their fragile economy is also at the mercy of world demand – when the aluminium price fell in the 1980s, a quarter of the Jamaican workforce lost their jobs and political unrest followed.

1 Why is cryolite added to bauxite before electrolysis?

2 Why are aluminium plants built near power stations?

3 Why is hydroelectric power used wherever possible?

4 A new aluminium plant is planned. You have been hired to suggest a suitable site. Write a letter to the owners suggesting the main factors to be considered. Try to recommend a suitable site.

6.11 The rarer metals

Element	% of crust
Silicon	26
Oxygen	50
Aluminium	7
Iron	4
Calcium	3
Potassium	2.5
Sodium	2.5
Magnesium	2
Hydrogen	1
Titanium	0.6
Carbon	0.3
Zinc	0.007
Copper	0.0045
Lead	0.0015
Tin	0.0002
Silver	0.000007
Gold	0.0000004

What on Earth?

The surface layers of the Earth contain 92 different elements, 70 of which are metals. Of these, iron and aluminium are quite common, but most of the other metals are much rarer. Though, in theory, it would be possible to get 5 milligrammes of gold (worth about 5p) from every tonne of granite, this would hardly be worth the effort or cost! So how do we extract these rarer metals economically?

Combining the elements

Most of the elements contained in the Earth are found as natural **compounds** which are mixed together in rocks. (These are the **minerals** – see 7.12). Looking at its composition, it is not too surprising that the simple compound of silicon and oxygen – **quartz** – is very common. Other common minerals such as felspar and mica are compounds of silicon and oxygen with aluminium. Iron is added to these, to form black minerals in igneous rocks.

Most metals form compounds too – typically with oxygen or sulphur – and these compounds are known as the mineral ores. These, too, are rare: but natural processes sometimes concentrate these ores so that we are able to extract them economically. How can this happen?

A small piece of ore from a gold-bearing vein.

Concentration from molten rock

During periods of mountain-building, deep within the Earth, huge volumes (tens of thousands of cubic kilometers) of rock melt to form **magma** – a kind of chemical soup! Some of this escapes to the surface as **lava**, but most remains to slowly cool and set deep underground . . .

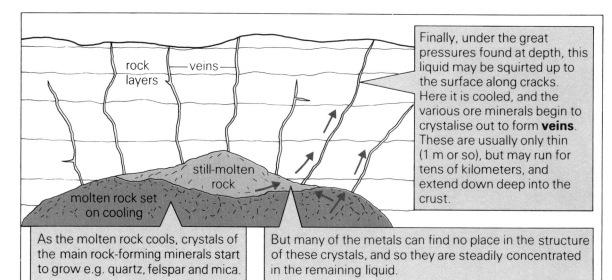

Finally, under the great pressures found at depth, this liquid may be squirted up to the surface along cracks. Here it is cooled, and the various ore minerals begin to crystalise out to form **veins**. These are usually only thin (1 m or so), but may run for tens of kilometers, and extend down deep into the crust.

As the molten rock cools, crystals of the main rock-forming minerals start to grow e.g. quartz, felspar and mica.

But many of the metals can find no place in the structure of these crystals, and so they are steadily concentrated in the remaining liquid.

Concentration by water

Ore veins may be weathered and eroded just like any other rock. Some ore minerals are hard and resistant, and remain as gravel or pebble-sized pieces at the bottom of rivers. But a pebble of ore may weigh two or three times as much as an ordinary pebble of the same size. Because of this, water currents may wash away the lighter material, leaving the ore in concentrated patches. Tin ore pebbles are often found at the bottom of Cornish rivers, and gold could be found in the sea-washed river sands of Sierra Leone.

Gold is one of the few elements to be found naturally pure and uncombined – 'native'. Because it is so dense, prospectors used to look for it in river beds, swirling the sand and gravel around in pans.

Now it's our turn

While ore minerals are themselves simple compounds, they are usually found as small pieces embedded in rock. Before we can extract the metal, we must separate these minerals from the surrounding rock. This is done in various ways. If the ore mineral pieces are large, the rocks may be *crushed* between giant **steel jaws** (this is often used as a first stage). If they are small, they must be finely *powdered*. This is often done in a '**ball mill**', where the rock is tumbled with steel balls. A third method is to use a '**hammer mill**', where the rock is beaten by flailing hammers!

*Sometimes several minerals may be present in an ore – this one contains zinc blende (brown), galena (grey), quartz (white) and . . . **fool's gold** (iron pyrites).*

Separating the mixture

It is now a matter of separating the mixture, and this is done in various ways. Many methods rely on the different densities of rock and ore mineral. Typical rock is about 2.6 times as dense as water, while copper ore is half as dense again. **Jigging** involves washing the mixture over shaking grids – the 'lighter' rock is washed clear of the denser ore, as in 'panning'. **Density separation** uses a dense slurry of iron oxide and water, the ore sinking to the bottom while the rock floats to the surface. **Froth floatation** works 'the other way'. Some powdered ores stick to the froth formed when air is blasted up through a sort of detergent. They rise to the surface and can be scooped off.

Now all that has to be done is to get the metal from its ore mineral!

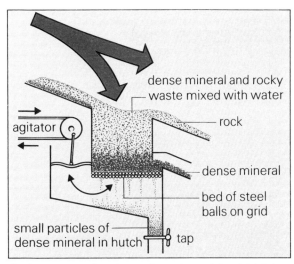

Jigging – separating the ore from the rock in which it is found.

1. What are the six commonest metals in the crust?

2. Why isn't it worth trying to get gold from granite?

3. How are the rarer metals concentrated
   a in molten rock;
   b by water?

4. How may ore minerals be
   a freed from rock,
   b separated out once free?

6.12 Metals and the environment

In the UK we live in a 'high consumption' society. We are constantly tearing at our planet to get the metals we need to keep up with the manufacture of goods. What effect does this have on our environment?

Mine it!

If an ore deposit has been discovered near the surface, the simplest way to get at it is to dig a big hole – which is just what was done when copper was discovered at Bingham Canyon, Utah, U.S.A. A hole was dug that covers 7 square km and goes down 800 m – and they're still digging! This method, called 'open-cast' mining, begins by stripping off the rock that covers the ore (the **overburden**) and dumping it elsewhere. The ore can then be blasted free and carried off for processing.

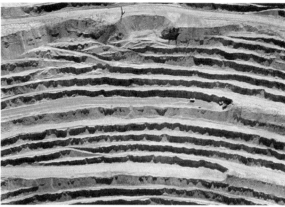
An open-cast mine – lorries carry the ore out along long winding roads cut into the side of the hole.

Reclaim it!

Mining companies like this opencast method of mining because it is relatively cheap, and miners like it because it is relatively safe. But conservationists hate it because it can ruin the environment. It is possible to minimise the damage, however, by carefully stripping and storing the topsoil at the start of mining. This can then be relaid after mining is complete, and replanted as forest, pasture or field.

In Britain, the law requires open-cast mines to be 'cleaned up' after use. This lake used to be a mine.

Refine it!

As you know, freshly mined ore still has to be further concentrated, and this generates vast amounts of 'useless' rock debris which has to be disposed of, often in ugly heaps. Yet even this might seem better than the sprawl of refineries and industrial complexes that produce and shape the metals we use.

It is only in relatively recent times that laws have been passed to control the pollution that once belched into the atmosphere from factory chimneys, or flooded into our rivers, killing fish and plants. Our atmosphere and our rivers may have been able to recover fairly quickly, but some areas where slag from furnaces was dumped decades ago remain as wasteland because of the poisonous metals that are still leaching out.

Refining metals can be a messy business – steelmaking factories need to be carefully run to avoid pollution.

Use or abuse it?

Some people say that our use of metals – in railway lines, in cars, and reinforcing the concrete buildings in our cities – has done nothing but harm to our environment. Other people say that all these things are great improvements – providing better communications, better housing for the growing population, and a better quality of life.

Certainly cars cause a particular type of 'metal pollution'. Lead is added to petrol to improve the way it 'burns' in the engine – but this lead ends up in the exhaust fumes, pouring out into the air we breathe. In some cities, this is so concentrated that it causes lead poisoning in children – enough to damage their mental development. Many countries are now moving towards lead-free petrol – but how many people do you know who use lead-free petrol?

Could you imagine a railway system without metal rails!

Motorway bridges need metal-reinforced concrete to support the weight of traffic without cracking.

Throw it away!

Between 10 and 20% of the metal produced every year is simply thrown away! We are all aware of how this can spoil the appearance of the environment, but it is more serious than that. Rusting metal can cut or injure children or wild animals, and chemicals spreading into the soil may stunt or kill vegetation. Zinc or lead-based batteries thrown away in or near ponds will eventually corrode and start to release their metals, causing environmental poisoning.

If we use the bins provided for our waste, and the services made available for disposal of wrecked cars and other 'technological debris', we can limit this damage to chosen dump sites, rather than spreading its menace across the countryside. But, better still, we could start to think of how to re-use our waste . . .

What a waste – of the metal and of the environment.

1 **a** Why do mining companies like 'open-cast' mining?
   b What are its drawbacks?

2 If an ore has to be concentrated by a factor of ten, how many tonnes of waste rock will you get for each tonne of 'purified' ore?

3 Find out how many garages in your area sell lead-free petrol. Do they still sell much more ordinary petrol?

4 Why were laws passed to control the chemicals that could be released into the air and rivers.

6.13 Spaceship Earth

▲ *There's no place like home – a view of the Earth from the Moon.*

Earth has often been likened to a spaceship. It is where we all live, and from it we must get all that we need. It may seem enormous to you, but there are nearly 4 billion people on the planet and we are consuming the Earth's resources at an ever increasing rate. You will have heard about the energy crisis and how our fossil fuels like coal and oil will soon run out. But that is also true of the metal ores, on which so much of our modern technology depends.

All 'mined out'

At present rates, and using today's mining techniques, many important metals will have been 'mined out' within the next 100 years or so!

Some optimists think that this is not really a problem. As our technology improves it becomes possible to extract metals from poorer quality ores. The waste tips from early mines have often been successfully reworked, and old mines – such as the tin mines of Cornwall – have been reopened when the price of the metal rose enough for them to show a profit again. (But they quickly closed again when the prices fell!)

But even then there are other problems to face. Some countries are rich in mineral resources, while others are poor. Industrial 'hi-tech' societies such as ours are using up far more than their fair share of the global resources, and many would say we are squandering them.

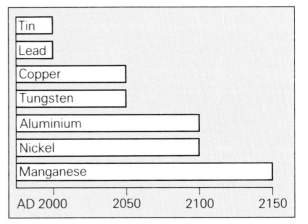

Going, going, gone! How much longer will metal ores last?

Should we ride 'first class' on spaceship Earth, while people in the developing world provide us with all we need? ▶

FACT·FACT·FACT·F
The foreign exchange earnings of Zambia and Chile were seriously depleted after the price of copper, the major revenue earner of both countries, dropped by half on international markets between 1974 and 1975.

Use and use again!

Much of the metal we use ends up on the scrap heap (see 6.12), but there is a better way. All we need to do is take a tip from mother nature...

Water is naturally recycled on the Earth – sea water evaporates, which makes clouds, which give rain, which runs in rivers back to the sea, and so on. The same 'raw material' – water, in this case – is used over and over again, and so can never 'run out'. If we could recycle all of the metals we use in the same way, our problem would be solved. This is happening to some extent. You've probably seen scrap metal yards near you – often full of rusting cars! We already recycle a significant proportion of some metals in the UK, as the table shows.

But there is still much to be done, and many problems to overcome. 'Pure' metals are easy enough to recycle, but many things today are made from alloys (see 6.3) – metal mixtures – from which the individual metals are hard to recover. Yet using expensive processes, some firms are now able to extract tiny amounts of precious metals such as gold and silver from scrap and still make a profit! Perhaps this is where 'improved technology' has the most to offer?

Where there's muck, there's brass – scrapyards make money by selling scrap metal for recycling.

% of metals recycled in Britain	
Aluminium	28
Copper	18
Iron	50
Lead	60
Tin	30
Zinc	30

Science fiction?

Of course, if we ever start to colonise space it would be very expensive to ship out raw materials for a moon base or a space station from Earth. One long term 'high tech' dream is to start to exploit the resources that exist out in the solar system. Many meteorites are found to be made almost entirely of iron and nickel – a mixture that exists deep in the Earth's core. It is thought that these might be the fragments of some broken planet, the bulk of which is now in orbit around the sun between Mars and Jupiter in the Asteroid Belt. Is it so far fetched to imagine that one day we might be able to mine these 'flying mountains' of metal?

Next stop – the asteroids?

1. Which metals are likely to run out in your lifetime?

2. Why do some people think that our metals will last longer than this?

3. In what way is our 'high tech' society damaging to other countries?

4. Describe how recycling helps us to conserve our metal stocks.

5. Write a science fiction story about 'mining the asteroids'.

6.14 Will we still need metals?

Maybe it's hard to imagine a world in which metals are not used – particularly after all that has been said in this topic. But, as technology advances, our needs change and other materials can be made to fit these needs.

Second best?

For a long time **plastic** was thought to be 'second best' – a cheap, inferior substitute for 'real' materials. But that is now no longer the case. Plastics are being 'tailor-made' to be the best materials now available for a whole range of uses. Many foods are now sold in **vacuum-sealed** plastic packaging instead of tin cans. This can cut down the cost, saves in weight and may even preserve the flavour better! Heavy and rust-ridden, cast-iron piping and guttering is being replaced by lightweight and 'rot-proof' polythene versions.

Why use plastic guttering? Because it won't rust. Its cheap and easy to carry up a ladder.

Getting better all the time

Many plastics may be further improved by reinforcing them with fibres of different kinds. **G**lass **R**einforced **P**lastic (GRP) has been in use for some time as lightweight yet strong construction material for canoes and playground novelties (among other things!).

Car manufacturers also use GRP for the bodies of some cars, to improve economy but not lose out on performance. A car built from GRP can give a 40% weight saving on a conventional steel version. This means that a smaller engine can give the same performance for a much lower fuel consumption!

A glass-reinforced plastic canoe, light enough to carry but strong enough to take the knocks.

Spacecraft operate in weightless conditions, but they must be made of lightweight materials to make it easier to get off the ground.

What's it worth?

The cost of weight saving in a car is only worth about £1 per kilogram, whereas for an aircraft the figure is nearer £50. So imagine what sort of costs might be saved in using plastics in spacecraft and equipment – where each kilogramme saved is worth more like £2000! It is hardly surprising then, that scientists' plans for space stations, orbiting 'factories' and solar power stations, should include the use of high-strength, low-weight, reinforced plastics wherever possible.

Metals can't do that!

One particularly important group of reinforced plastics (composites) uses carbon fibres. Whereas metals are more or less equally strong in any direction, carbon fibres are very strong along their length, but quite weak at right angles to this. This disadvantage can be made up for by matting the fibres in different directions, giving a light but very strong material which is being used increasingly in 'high performance' aircraft.

But this 'weakness' can also be turned to advantage. With the aid of computer design, the fibre matting can be arranged so as to make the finished plastic stiff and strong in some directions, while being flexible in others.

The flexibility of plastics reinforced with carbon fibre is used in the tail rotors of modern helicopters. The rotors flex in use, removing the need for complex systems of bearings.

The same 'flexibility' has also been used in the design of advanced aircraft wings, allowing greater manoeuvrability in flight.

Lasers in action

In today's telephone systems another important change is taking place. Until recently, copper cables were used to carry the electrical signals. But the copper was expensive, the cables big and awkward and sometimes electromagnetic intereference made conversation difficult. Now this is all changing. The messages are now carried by laser light along thin optical fibres. These are made of high quality glass (and therefore made from sand!) that is so pure that a window 1 km thick would seem as clear as an ordinary pane. This high quality glass is then coated with a layer of different glass which traps the light in the fibre.

These optical fibres are very thin – just one twentieth of a millimetre in diameter yet, because of the way the laser technology works, a single pair can carry 2000 telephone calls at the same time!

Plan and conduct a survey to find out the extent to which plastics or other materials are beginning to replace metals in your school, home and local area. Discuss the implications of your findings.

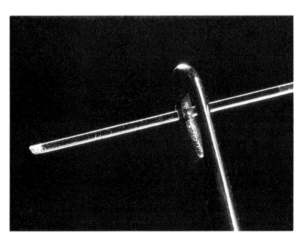

Thick copper cables are now being replaced by very thin optical fibre bundles, which leave plenty of space for expanding the system.

MODULE 6 METALS

Index
(refers to spread numbers)

A
acids 6.5
alloys 6.2, 6.3
aluminium 6.1, 6.2, 6.10
amalgam 6.3
annealing 6.4
anode 6.10

B
ball mill 6.11
bauxite 6.8, 6.10
blast furnace 6.8, 6.10
batteries 6.1
brass 6.3
bronze 6.3

C
carbon 6.4, 6.8
cast iron 6.4
cathode 6.8, 6.10
coke 6.9
compounds 6.5, 6.11
copper 6.1, 6.2, 6.5
corrosion 6.6.6
cryolite 6.10
cupronickel 6.3

D
density 6.2
    separation 6.11
duralumin 6.3

E
electrolysis 6.8
electrical energy 6.1

F
froth flotation 6.11

G
galena 6.8
German silver 6.3
gold 6.2
GRP 6.14
gun metal 6.3

H
hammer mill 6.11
high carbon steel 6.4

I
iron 6.2, 6.4, 6.6, 6.9
    cast 6.5
    wrought 6.5

L
lasers 6.14
lava 6.11
lead 6.1, 6.2
limestone 6.9

M
magma 6.11
magnesium 6.2
manganese 6.13
medium steel 6.4
mercury 6.2
metal ores 6.8, 6.11
meteorites 6.13
mild steel 6.4
minerals 6.11
mining 6.7
molten metal 6.9

N
nickel 6.9

O
overburden 6.12
oxide layer 6.6

P
paint 6.1
plastics 6.14
platinum 6.1

Q
quartz 6.11
quenching (of metals) 6.4

R
reactivity 6.5
    order of 6.5
reclaiming (mines) 6.12
reduction reaction 6.8
refining 6.12
rusting 6.6

S
scrap metal 6.13
'self sealing' metals 6.6
silver 6.1, 6.4
slag 6.9
sodium 6.5
steel 6.2, 6.4
strength 6.2

T
tempering (of metals) 6.4
titanium 6.2, 6.7
tungsten 6.2

U
undersealed 6.6

V
vacuum sealed 6.14
veins (of ore) 6.7, 6.11

Z
zinc 6.4, 6.6

Photo Acknowledgements

The references indicate the spread number and, where appropriate, the photo sequence.

Aviation Picture Library (Austin J Brown) 6.3/4; Barnaby's (Sue Barnes) 6.1/3, David Kirby 6.1/4, (Roy Perring) 6.4/1, 6.12/4, 6.12/5; British Coal 6.12/2; J Allan Cash – contents 6.3/2, 6.3/3, 6.3/5, 6.4/1, 6.12/4, 6.13/3, CEGB 6.1/1, Norman Childs Photography 6.2/4; Ellawayday Pics 6.2/1; Ford motor company 6.6 main pic; Sally & Richard Greenhill 6.5/1; GeoScience Features 6.4/4, 6.9/1, 6.11/1, 6.11/2, 6.12/1; Trevor J Hill 6.13/6, 6.14/1; Michael Holford 6.2/3, 6.3/1, 6.8/1, 6.8/2; Hutchinson (Pru Rankin Smith) 6.10/2; Colin Johnson 6.4/2, 6.5/2, 6.5/3; Frank Lane Agency (R P Lawrence) 6.12/6; Military Archive & Research Services 6.7/1, 6.14/4, 6.14/5; Science Photo Library (Ben Johnson) 6.2/2, (Alexander Tsiaris) 6.7/2, 6.12/3, (NASA) 6.13/1, 6.13/4, 6.14/3; Sporting Pictures 6.8/3, 6.14/3; Topham Picture Library 6.9/3; Debbie Walker 6.1/2

Picture research : Jennifer Johnson

MODULAR SCIENCE for GCSE

MODULE 7 Earth Science

The **Earth** is our home and the source of all our materials. This module looks at the structure and features of the Earth, and how these change over time.

Relevant National Curriculum Attainment Targets: 9, (5), (7)

7.1	Our restless Earth
7.2	Earth – the inside story
7.3	Volcanoes
7.4	Earthquakes
7.5	Plate Tectonics
7.6	The rock cycle
7.7	Breaking rocks . . .
7.8	Chemical attack
7.9	River and sea
7.10	Finding out about the past
7.11	How time passes!
7.12	Earth's mineral resources
7.13	How to identify minerals
7.14	Armero – an avoidable tragedy?

Index/Acknowledgements

MODULE 7

7.1 Our restless Earth

'Looks harmless enough'

Imagine that you are in a spacecraft – looking out your window you would be able to see the whole of the Earth. You would probably see white swirls of cloud moving over either green and brown land or the blue seas. The Earth might look peaceful at this distance, but it is not always such an attractive place . . .

4000 DEAD IN MEXICO CITY

THE public clock outside the Credito Mexicano bank on the Paseo de la Reforma in the centre of Mexico City stopped after two minutes of the earthquake at 7.22 a.m. By that time the old Hotel Regis next door, one of the city's most notable landmarks, had collapsed into a pile of rubble crushing in its masonry a full complement of guests enjoying the height of Mexico's tourist season.

Marta Anaya, a reporter on the *Excelsior* newspaper, remembers the moment the city shook: 'First came silence, fear and uncertainty. Then there was the shaking that didn't stop. The lights wobbling, the windows breaking, the floor quivering. And later the innards of the earth seemed to crumble. Buildings collapsed, houses fell, schools disintegrated and the people shouted, cried, implored. . .'

'A dangerous place to live'

■ Scientists believe the devastating Mexican **earthquake** was triggered as two giant rock **plates** met and slid past each other deep underground.

The release of energy caused one of modern history's most destructive quakes, comparable only with those in 1906 in San Francisco and in 1976 in Tangsnan, China.

It produced a reading of nearly eight on the **Richter scale** of tremors, one of the highest ever recorded.

'This is not surprising,' said Dr Peter Smith, reader in Earth Sciences at the Open University. 'Mexico lies at the centre of a complex mish-mash of geological forces. It can be a very destructive and dangerous place to live.

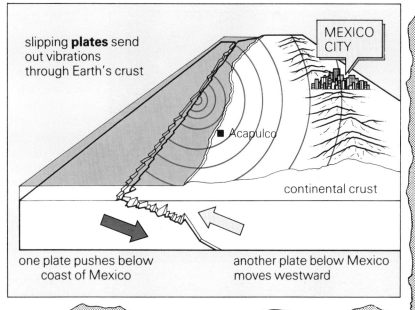

What on Earth is going on?

We read about terrible disasters such as this quite often. In many parts of the world, people live with the risk of their homes and lives being devastated by earthquake or volcanic eruption. These great forces are so powerful that they reshape the surface of the Earth. Yet these forces do not just destroy: over millions of years they have built the huge areas of land that cover one-third of the Earth. We depend on this land for most of the food which keeps us alive – and for many of the natural resources we use in our everyday lives.

So we need to find out more about our restless Earth – that will help us get what we need from it. Also if we can learn to predict when and where disaster will strike, we may also be able to save many lives.

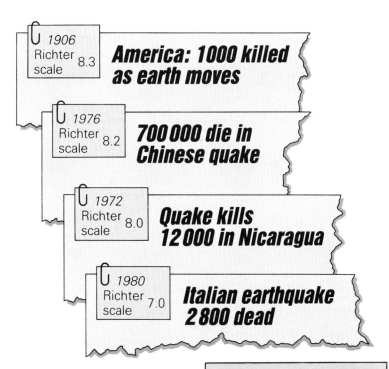

'Quakes have always been a threat – but crowded modern cities make them more deadly than before.

The special area of scientific study which deals with these problems is known as **Earth Science**. What's so special about Earth Science? You will find out in this module. ▶

7.2 Earth – the inside story

How can we tell?

The Earth is a planet – a great ball of rock 13 000 km across! How can we hope to find out what it's like inside? Our deepest mines scratch barely 13 km into the surface. The answer is to use the same trick that builders use to find out what's behind a wall – tap it and listen to the vibrations! Of course, it's not quite that simple. You would need a very loud tap to pass right through the Earth. But nature often does the job for us – with earthquakes!

Measuring earthquakes

We can easily feel earthquakes if we are unfortunate enough to be standing too close. But we need to use a special instrument to measure the tiny vibrations from distant ones.

This instrument is called a **seismometer**, and the world is now dotted with seismometer stations. These are able to detect even the tiniest tremor. The results are then put together and analysed by computers.

What do they tell us?

Every earthquake sends shock waves out across the surface of the Earth. These cause the damage, but also help to pinpoint where that damage is likely to be worst – at the **epicentre**. This is the point directly above where the rocks moved.

Seismometers can't tell where the shock waves have come from, but they can tell how far they have come. Drawing distance circles from three stations fixes the epicentre and so helps to direct aid to the worst affected areas.

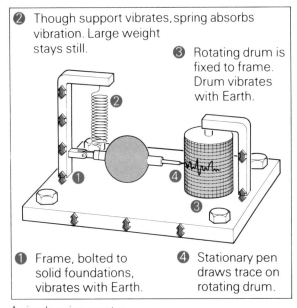

A simple seismometer.

A seismometer recording.

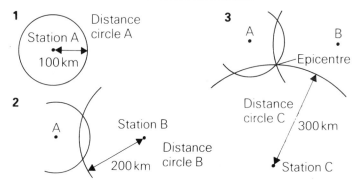

Three distance circles will locate a 'quake epicentre precisely.

A soft boiled Earth

But the shock waves also pass down into the Earth. It is these fainter signals that can give us an 'inside-view' of our planet. The results tell us that we live on a round, rock version of a soft-boiled egg!

The crust The hard, brittle outer layer is known as the **crust**. Compared to the diameter of the Earth, the crust is very thin. You could compare its thickness to that of a piece of 'Cling-film' wrapped around a football. It is broken in places, but still holds together – rather like the cracked shell of a soft-boiled egg.

The mantle Beneath the crust is a very large layer of dense, hot, semi-liquid rock – the **mantle**. Over millions of years, the mantle moves around. This gradually changes the shape of the crust above the mantle. In the terms of the 'softboiled egg' model, the mantle is like the white of an under-cooked boiled egg.

The core The Earth has a liquid centre – the **core** – which consists of molten metal (iron and nickel). Radioactive decay in the core and mantle is the source of all the heat energy released by the Earth. The core is like the yolk of an egg, but with strong swirling movements that keep the core well-mixed.

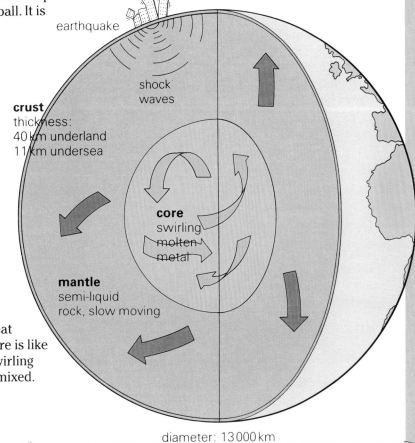

How did it happen?

5000 million years ago, the Earth was a molten ball of rock and metal. On cooling the **crust** formed – just like ice forming on the surface of water (or the skin forming on custard). At the same time, the dense iron and nickel sank down to form the **core**. This left behind the molten rock which formed the **mantle**.

1 How can we find out what the inside of the Earth is like?

2 Describe how a seismometer works.

3 How can shock waves be used to direct aid to earthquake disasters?

4 Describe simply the layers of the Earth. Do these layers remind you of anything apart from an egg?

5 **a** How did the crust of the Earth form?
b What causes part of the crust to change shape?

7.3 Volcanoes

A volcano is born

In 1943, a Mexican farmer noticed that foul-smelling gases were escaping from a crack in the ground in his field. His surprise soon turned to alarm as gas pressure gradually tore open the hole. Fragments of rock were thrown out, forming a mound of rubble. By the end of the day, molten rock – **lava** – began to pour out from the crack. This lava cooled to form a mound-shaped cone. Within 10 years the new volcano, called Paracutin, was 400 metres high!

A volcano can look pretty spectacular – but the effects can be very dangerous.

What happened?

Volcanoes form when molten rocks (from far below the crust) force their way up to the surface of the Earth.

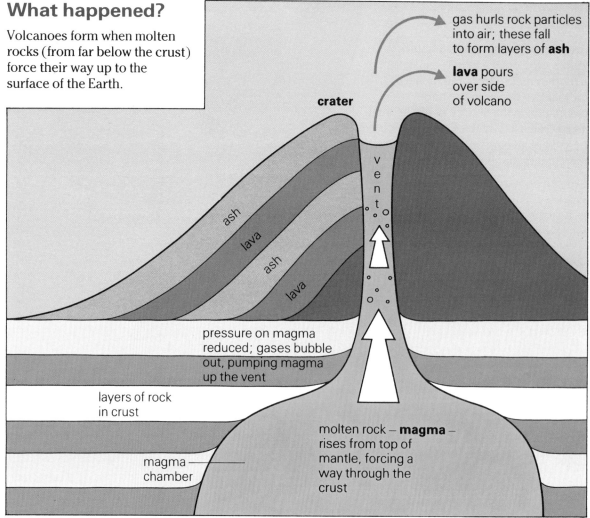

An explosive story

Some volcanoes have craters which stay open between eruptions. Others, such as Vesuvius, are of a 'self sealing' type – the lava in the crater sets to solid rock between the eruptions. Because of this, gas pressure within the volcano cannot be released and so it steadily builds up. It becomes a giant time bomb. Eventually the pressure will become so great that the top of the volcano will simply be blown off – like a bottle of Champagne blowing its cork.

During Roman times, Vesuvius was thought to be **extinct** (completely inactive) and the city of Pompeii grew at its side. In 79 AD, however, the lava plug could no longer contain the enormous pressures within and a violent **eruption** occurred. With very little warning, the people of Pompeii were showered with red hot ash and were killed where they stood. The ash completely buried the city. Pompeii remained buried until the layers of ash were removed in the last century.

Suffocated by ash, this Pompeii citizen was soon buried by more falling ash which formed a permanent cast of the body.

Why live near a volcano?

The people of Pompeii had an excuse – they didn't realise that Vesuvius was still an **active** volcano. But today, people often live close to volcanoes that they know could erupt at any time. Why do they take such risks? The answer lies in the soil. Lava forms a rich, fertile soil which is very good for growing large amounts of food.

Free heat

The hot rocks beneath a volcano can be used as a source of heat energy. A geothermal power plant can convert this heat energy into electricity. These power plants pump water down to the hot rocks. Here the water turns to steam, which expands and forces its way back to the surface. The steam then turns a turbine and electricity is generated in the usual way (see 4.13). The advantage of this method is that there is no pollution. In some places, the hot water is even pumped direct to homes – providing free hot water!

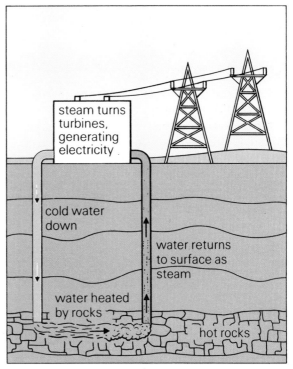

You can't get blood out of a stone – but you can get electricity out of hot rocks!

1 Describe what lies below a volcano.

2 What causes molten rock to move up a volcano vent?

3 Give **two** advantages of living near a volcano.

4 What costs will be involved in running a geothermal plant?

5 Write an 'eye-witness' account of the destruction of Pompeii.

7.4 Earthquakes

Disaster!

Every few months, news reaches us of some terrible natural disaster where a powerful **earthquake** has struck, destroying buildings, killing many people and making thousands homeless. Reports tell of deep rumbling noises which caused animals to flee, followed by a great shaking of the ground. It is as if the land were a great carpet which was being shaken out – throwing the land up in waves and shattering buildings. Often, great gashes appear in the earth.

But not here . . .

Fortunately, you will never experience anything like that in Britain – unless you visit the 'earthquake room' at the Geological Museum in London! Earthquakes are so rare here that even minor tremors hit the headlines.

'Safe as houses'? Not during a 'quake!

DAILY MIRROR, Friday, July 20, 1984

Britain rocked by biggest tremor

BRITAIN was shaken out of bed yesterday by its biggest-ever earthquake.

The country was rocked from end to end by a tremor centred near Portmadoc in North Wales.

The quake measured 5.7 on the Richter scale, higher than the "Great Victorian Earthquake" which rocked Colchester 100 years ago.

Here's how yesterday's quake had Britain all a-quiver:

In LIVERPOOL, chimney pots toppled from several houses and two elderly ladies were treated for shock when one crashed through their roof.

In NORTH WALES, windows were shattered and power failures were reported near Pwllheli.

The worst we can expect in Britain.

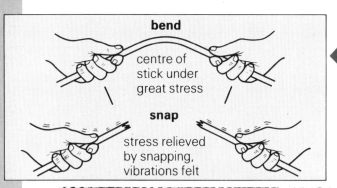

bend — centre of stick under great stress

snap — stress relieved by snapping, vibrations felt

So what is an earthquake?

Think what happens when you snap a dry stick. At first it bends, but eventually it snaps suddenly, sending shock waves back along your arms. A similar thing is happening in the Earth. In places, the layers of rock are put under great **stress** and eventually, like the stick, they snap. This sends shock waves out through the Earth's crust – causing an earthquake.

California, here I come?

Britain is in a very stable part of the Earth's crust now, but California is not so lucky. San Francisco is built across an enormous crack in the Earth's crust called the **San Andreas fault**. Two massive blocks of rock are moving sideways along this crack at about 7 cm a year – look how this has shifted the orange trees in the photograph. Around San Francisco the blocks have jammed and so pressure is steadily building up as it did in the bent 'stick' experiment! Would you like to live in San Francisco?

These orange trees were planted in straight lines – then the San Andreas fault moved!

The scale of disaster

Thousands of earthquakes are detected every year, but only a few of these are severe enough to cause damage. Their strength is measured against the **Richter scale** – the higher the number, the greater the shock!

The Portmadoc earthquake measured just 5.7 on the scale – almost an 'everyday occurrence' in California! In 1906, San Francisco was completely detroyed by an earthquake estimated as having been between 8 and 9 on the scale.

Richter scale	Effect of 'quake
3	only shown by seismometers
4	slight – like heavy traffic
5	quite strong – bells ring, sleepers wake
6	very strong – walls crack, plaster falls
7	destructive – some houses collapse, ground cracks
8	disastrous – few buildings left, landslides and floods
9	catastrophic – total destruction, ground thrown into waves

Does it help?

Scientists record details of all earthquakes, large and small. They search for patterns that might help to predict where disaster will strike next. This worked successfully in 1980 when a sudden burst of small tremors was detected in the Cascade mountains of the USA. The scientists warned people to evacuate the area before Mount St Helens erupted. This volcano had not erupted for 120 years, so none of the locals had suspected the danger!

Scientists now know that minor shocks often come before major earthquakes, so they are listening to the area round the San Andreas fault very carefully!

Building for safety

Regardless of risk, millions of people live in danger from earthquakes. What can be done to protect them? For centuries the Japanese built lightweight bamboo and paper houses, but these are not suitable for a modern city.

After the Mexico city 'quake, many people were surpised to find tall skyscrapers still standing, while older brick buildings had shattered and collapsed. That was because the reinforced concrete framework was flexible enough to sway with the vibrations! This idea is used in the most modern earthquake-proof buildings.

Buildings which are narrow at the top are less likely to be damaged during a 'quake.

1 What causes an earthquake?

2 Why are the people of San Francisco so concerned about earthquakes?

3 Draw a Richter scale which shows the damage in pictures.

4 Why do scientists measure and record every earthquake?

5 Would you rather be in a skyscraper or a large brick house in an earthquake? Why?

6 Write a story about being caught in a earthquake.

7.5 Plate tectonics

The heat is on

Have you ever noticed how warm air rises above a radiator, while cold air flows out over your toes when you open a fridge? This also happens when you heat water: the hot water rises, and a swirling **convection cell** is set up (see 4.7).

Something very similar seems to be going on in the rocks of the Earth's mantle. These rocks seem fairly solid, yet over millions of years, they can change shape and circulate round just like water! This has caused the Earth's crust (and a slab of mantle) to crack up into a gigantic jigsaw-puzzle of pieces called **plates**. And what is more, these plates are still moving around . . . by a few centimetres every year! That is enough to cause earthquakes where the plate edges rub against each other. Over millions of years, this process can make major changes to the surface of the Earth. This process of change is called **plate tectonics**.

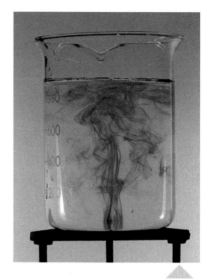

Convection: The purple colour rises with the hot water at the centre. As the water cools, you can see that it moves to the side and then downwards.

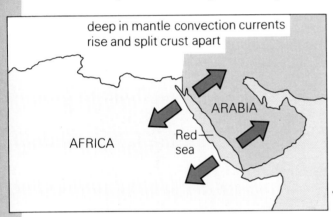

Cracking up!

Where hot mantle rock rises, it splits the crust apart! This is even happening today – in Africa. Arabia continues to be torn out of Africa, opening up the Red Sea by a few centimetres every year.

300 million years ago, Europe was joined to America: but a large crack formed. Molten rock rose up from the mantle to fill the gap. This crack has continued to widen ever since. Topped up with water, the result is the Atlantic ocean! Iceland sits across the crack – and is getting wider every year!

Fold mountains

If some of the jigsaw pieces of the Earth are moving apart, others must be coming together! When this happens the sediments at the edges of the plates are folded and crumpled into new mountains. Explosive volcanoes are common in these areas, such as those in the '*Ring of Fire*' around the Pacific.

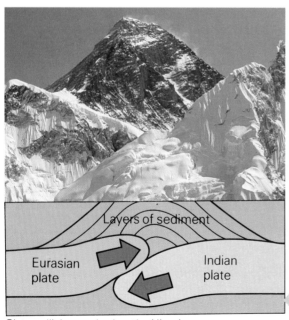

Plate collision pushed up the Himalayas.

Where continents collide, great **fold mountains** form between them. The Alps formed as Africa collided with Europe, and the Himalayas rose up as India rammed into Asia.

The plate tectonic jigsaw

If we look at the way the plates of the Earth are arranged at this moment in time, and at the way they are moving, it is easy to explain why some areas are more at risk than others...

Britain is at the centre of the Eurasian plate, more than 1000 km from the edge. This means there are no active volcanoes and no severe earthquakes.

A belt of fold mountains runs from the Alps to the Himalayas. These mountains are 'young' – less than 100 million years old – so these areas still have violent earthquakes and volcanoes.

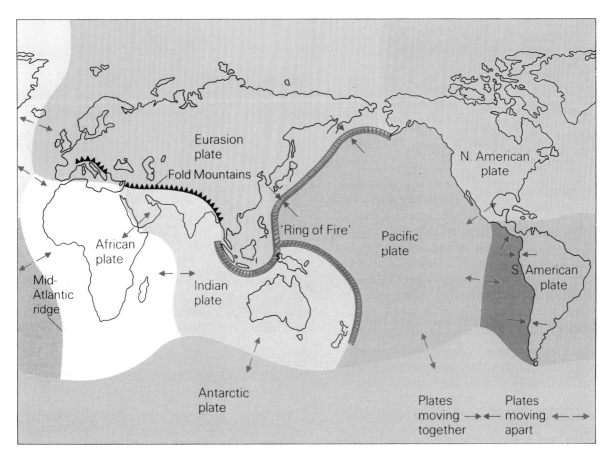

Down the centre of the Atlantic there is a line of undersea volcanoes – the **mid-Atlantic ridge**. This line marks where the plates are cracking apart. There are similar ridges in the Indian and Pacific Oceans.

All around the Pacific Ocean, plates are colliding. This has formed a '*Ring of Fire*' – chains of explosively volcanic islands. Severe earthquakes are very common, often causing destructive tidal waves.

1 What parts of the plates cause earthquakes?

2 What process makes the plates move?

3 Why is Africa 'cracking up'?

4 How did the Himalayas form?

5 Why is Iceland getting wider?

6 **a** What is the 'Ring of Fire'?
   b Suggest other areas that might have a lot of volcanoes.

7.6 The rock cycle

Let's look at rocks...

We have mentioned rocks quite a lot so far, which is not surprising as the whole of the Earth's crust is made of rock. But what exactly are rocks? Scientists talk of three main types – **igneous**, **sedimentary** and **metamorphic** – which form in very different ways, yet are linked together in the Earth's **rock cycle**...

Granite – an igneous rock with large crystals formed by slow cooling.

Igneous rocks

There are many types of **igneous** rock, but they all have one thing in common – they were all once molten **magma** that rose up from the base of the crust. As the magma cooled, *crystals* started to grow from the liquid. Because of this, the crystals in the rock are formed in a random way.

You can recognise an igneous rock by the fact that it is a hard rock made of tightly interlocking crystals which are scattered randomly. This crystal formation is easy to see in **granite**, which formed by cooling and setting *slowly*, deep underground. But you would need a microscope to see it in **lava** which forms by cooling and setting *rapidly* on the surface of the Earth.

Basalt – an igneous rock with small crystals formed by fast cooling (magnified x 10 approx.).

Sedimentary rocks

Sedimentary rocks are formed from the broken fragments of other rocks, which were washed down to lakes or seas by rivers, where they collected as *layers* of **sediment**. These were then squashed and hardened into rock.

You can recognise a sedimentary rock by the fact that it is made of broken pieces – pebbles, sand or clay – rather than crystals. These pieces are often stuck together by a kind of natural cement, but may still be quite soft.

Conglomerate – a sedimentary rock made up of pebbles and natural cement.

Metamorphic rocks

Sedimentary rocks often get caught up in the collisions of plates, and are forced up to form fold mountains. Deep below the mountains, the rocks get incredibly *hot* and are also *squeezed* by immense pressures. Under these conditions, the rocks start to recrystallise without actually melting. This makes the crystals grow in definite bands or layers, forming **metamorphic** rocks.

You can recognise a metamorphic rock by the fact that it is made of crystals which are arranged in bands or layers.

If metamorphic rocks get so hot that they melt, then igneous rocks will be formed. The rock cycle has started again!

Gneiss – a metamorphic rock with banded crystals.

All change!

Over a very long period of time, each type of rock can be changed into another type...

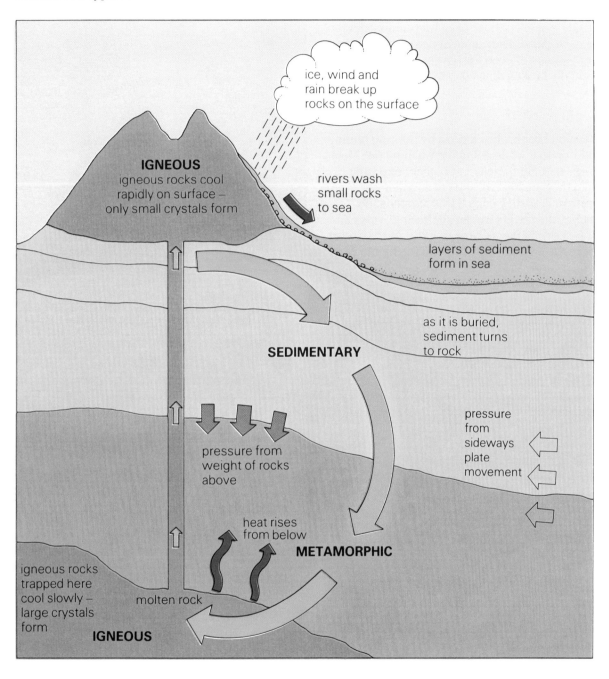

1 What do all igneous rocks have in common?	*4* What is the main visible difference between igneous and sedimentary rocks?
2 Explain the difference between granite and lava.	*5* How do metamorphic differ from igneous rocks?
3 How do sedimentary rocks form?	*6* Describe the rock cycle.

7.7 Breaking rocks...

Look at the rocks in the mountains or at the seaside – or even those used for buildings in the towns or cities. They often appear hard and everlasting. Yet even the hardest rocks are gradually broken down or worn away...

...by heat

Have you ever broken a glass by pouring boiling water on to it while doing the washing up? Most things, including rocks, get bigger (**expand**) when they are heated, and get smaller (**contract**) as they cool. If this happens too quickly, the results can be shattering! This can happen over a longer time to rocks in the desert, where the days are very hot and the nights very cold.

...by cold

A more common cause of breakage is ice-shattering. Unlike most things, **water expands when it freezes**. The force of this is great enough to burst a water pipe – or a rock!

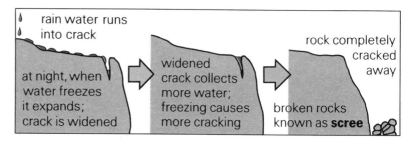

rain water runs into crack

at night, when water freezes it expands; crack is widened

widened crack collects more water; freezing causes more cracking

rock completely cracked away

broken rocks known as **scree**

...by plants

Tree roots often force their way into the cracks in rocks. Over the months and years, the roots split the rock apart.

Tree roots can easily split rocks – and often break up concrete drives.

This is a very destructive way to obtain samples.

...by us!

Humans and other animals often speed the break-up of rocks. Well-worn paths, mining, and even scientific investigation, all cause rocks to break up.

Pebble rolling

Heavy rain will eventually wash these fragments of rock into rivers, where they will be rolled and tumbled by the current. As they knock against each other and the river bed, they are steadily worn away. The corners get chipped off first. So the further they travel, the rounder they become – as well as getting smaller. This process is continued by wave action in the sea, and beach pebbles are often very well rounded.

Gorges and valleys

Fast-flowing rivers roll pebbles along the river bed. These pebbles bump into each other – but they also bump into the rocks over which they are rolling. In this way, the rocks beneath a river are ground away – **eroded** – and the river cuts down through them. The eroded rocks form a fine sediment that gets carried down to the sea.

If the rocks are hard, a steep-sided **gorge** may form. But normally the sides collapse under their own weight, leaving a typical **'V' shape** of valleys. This may be quite steep-sided in the mountains, but flattens out downstream as the water flows more slowly.

Ice and wind

Rocks can also be broken down and carried away by the *movement* of ice and wind.

This gorge has been made by a fast flowing river cutting through the rock.

Glaciers are huge moving blocks of ice that can carry piles of broken rock (known as moraine). If caught at the glacier base, boulders gouge out the ground to form U-shaped glacier valleys.

The wind carries sand through the air, blasting rocks in its path. The rocks are usually worn away just above ground level.

1 Name two ways that ice breaks up rocks.

2 Why is it important to fill in holes in brickwork?

3 In what ways are broken rocks used as 'tools' to cause more rocks to be eroded?

4 Why are pebbles round?

7.8 Chemical attack

Weathering away!

Most rocks can be *physically* broken down by the weathering effect of wind, rain and so on (see 7.7). Certain rocks also come under **chemical attack** from the *air* and *water*. In the case of igneous rocks, chemical weathering is the main cause of their break-up. Why should this be so?

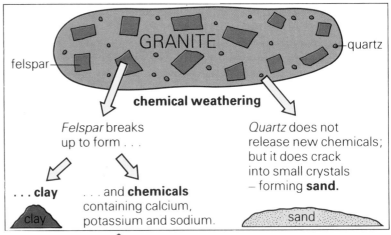

Chemical weathering causes granite to break-up.

Igneous rocks, such as granite, were first formed in fierce conditions deep underground. When they reach the surface, the physical attack by the wind and rain is only a minor threat, compared to the heat and pressure underground. However, when deep in the crust, igneous rocks did not suffer from chemical attack by air and water. Although they can easily resist physical weathering, once igneous rocks are exposed on the surface they will be (slowly) broken down by **chemical weathering** to form clay.

Under tropical conditions, this chemical weathering goes a stage further – it breaks down clay to form **bauxite** (the raw material needed for aluminium production see 6.10).

Soil

Left undisturbed, the processes of physical and chemical weathering produce a layer of broken and decomposed material over the rocks of the Earth. We call this layer the **soil**. Plants grow readily in it, as it contains all the minerals they need – and it traps the water they need as well!

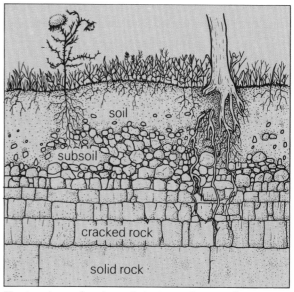

The break-up of rocks and plants produces soil.

Disappearing soil

Soil is often very loose and can easily be washed or blown away by the wind or rain. It is only held in place by the roots of plants and trees. In many tropical countries, patches of **rainforest** are being cleared to grow crops and to sell the timber. But without the **binding action** of the trees' roots, the soil is quickly washed away. Vast areas have lost their soil in this way and have become useless for farming.

Cleared rainforest soon loses its soil – and its usefulness.

Limestone under attack

Limestone is a special kind of rock that was formed millions of years ago from the **shells** of dead sea animals. These shells collected on the sea bed and were gradually covered by thick layers of sand and mud. Eventually the shells were changed into limestone – some examples are tough and grey, others are soft and white (chalk). It can even be metamorphosed into marble. Whichever form it takes, the rock is made of **calcium carbonate** (lime). This means that limestone rocks are at risk from chemical weathering.

Calcium carbonate reacts with a group of chemicals called acids. Rainwater is a **weak acid** – this is because it contains **carbon dioxide**. Carbon dioxide is a gas found in the air. As the drops of rain fall through the air, carbon dioxide **dissolves** into the water droplets. This is what causes the rainwater to become acidic. When the slightly acidic rainwater falls onto limestone rock, the rainwater reacts with the calcium carbonate in the rock. This chemical reaction causes the calcium carbonate to dissolve in the rainwater. The limestone rock then breaks up – a bit like when a sugar-cube breaks up when put into a cup of tea only much slower!

Limestone is dissolved by rainwater – this causes cracks to form in limestone statues.

Caves in limestone

Limestone has natural cracks which are easily attacked and widened by the acid in rain. Once these have opened up enough, whole rivers can disappear into the rock down '**swallow holes**'. The rivers then flow underground, widening the cracks into tunnels and underground caves.

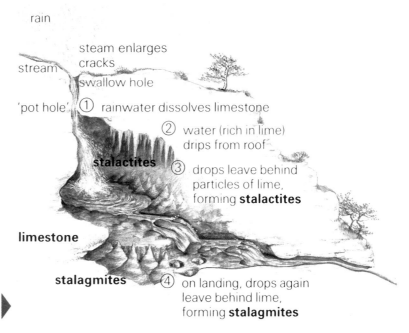

The formation of limestone caves.

1 What causes the break-up of igneous rocks?

2 What is soil? Why is it important to plants?

3 Why are plants important to the soil?

4 Why does limestone weather away so quickly?

5 How do stalactites form?

6 A timber company is planning to cut down a rainforest. They say that this will also help to create new farmland. Write a letter to them, explaining your views of their plans.

7.9 River and sea

River action

When rain falls in the mountains, some rainwater will soak into the ground – but most will 'run off' the hillsides. This water forms streams, which join to form rivers, and so on – all the way to the sea. Rivers **erode** (wear away) the rocks over which they pass, and carry the debris away. The amount of debris the river can carry depends on how fast it is flowing. More debris, or **sediment**, may be carried in one short period of 'flood' than in the rest of the year put together!

River deep!

A river in flood carries a vast amount of sand and mud. However, if it 'breaks its banks', the water spreads out and slows down, and so cannot carry the sediment any more. This is then dumped at the side of the river, forming a flat **flood plain**. The sediment contains many chemicals which help plants to grow – this makes the flood plain ideal for farming. A temporary flooding of the fields may be a nuisance, but when the flood waters drain back into the river, rich, fertile soil is left.

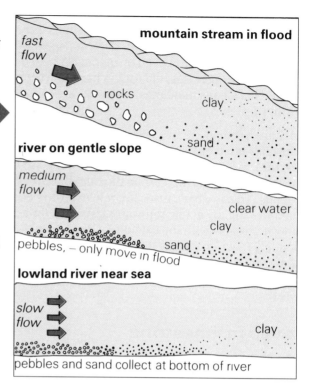

The sea continues the break-up of pebbles that was started by the river.

Flooding fertilises fields!

The Nile delta – seen from the Space Shuttle.

The sea at last!

Most of the sediment carried by a river will eventually reach the sea, and here the river's power is lost and the river dumps its load of sediment. Now the sea takes over – wave action stirs and sorts out the sediment: fine clay is washed out to sea by even the weakest of currents, but any sand or pebbles are spread out along the beach.

If a very large amount of sediment is brought to the sea by a river, the sea may not be able to move it all. A great wedge of sand and mud (a **delta**) may build out into the sea as new land. The Nile delta is so big that it is easily seen from space. It is an important farming area in Egypt, where so much of the rest of the land is desert – no good for growing crops.

Layers on the sea bed

Eventually, most of the pebbles, sand and clay brought in by the rivers end up in more or less horizontal layers – **beds** – at the bottom of the sea. These are separated by **bedding planes** – natural breaks that were once the seabed, and over which sea creatures may have left their scattered remains.

In time, these sediments will be buried and squashed by other layers which form above them. The sediments finally harden to form sedimentary rocks – and the shells in the sediments are changed into **fossils**.

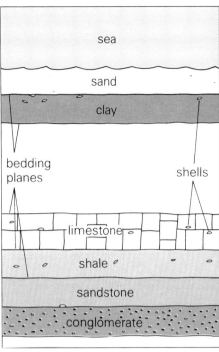

Sediment layers underneath the sea-bed.

*Pebbles form **conglomerate** rock – this is a bit like concrete, with the pebbles held together by a natural cement.*

Sand forms **sandstone** – usually soft, but some old sandstones can be very hard.

Squashed mud forms **shale** – this is flaky and often contains fossils. The layers have been twisted up by movements in the crust.

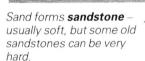

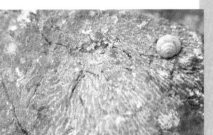

Crushed shells form **limestone** – this also contains many fossils, such as the one shown here.

What else comes along

As well as sediment, rivers bring many chemicals to the sea in **solution**. These chemicals were dissolved from the weathered rocks by rainwater. Like the sediment, these chemicals remain in the sea, which is why the sea is salty!

Lots of calcium (dissolved by rain from limestone) is brought by the rivers to the sea but most is taken by shellfish in order to make their shells. Given enough time, these shells will one day form new limestone rocks.

1 What controls the size of the particles that a river can carry?

2 Why can floods be helpful as well as harmful?

3 Water flows only slowly round the inside of river bends. Predict what will happen to the sediment carried by the river in such places.

4 Why is the sea salty? Where does the salt come from?

7.10 Finding out about the past

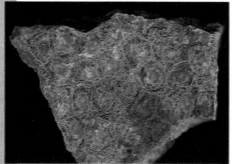

How many different faults can you see?

The rocks around us

We are lucky in Britain to live in such a stable area – free from volcanoes and earthquakes. Yet looking around our coasts and mountains, we can see much evidence that this was not always so. Layered sediments have been **folded** and **faulted**, and igneous and metamorphic rocks are common in some areas. All of this is *evidence* which helps us to understand what must have happened in the past.

Firstly, we can compare what we see with what we *know* happens today: pebbles collect on beaches; lava comes out of volcanoes; faulting occurs when blocks of rock are torn apart during an earthquake – and so on.

Secondly, we can get some idea of the *order* of past events from the order of sedimentary rocks. The oldest layers are those at the bottom; older rocks might also be more altered, or might have suffered more folding and faulting, and so on.

Any old bones?

Fossils are often found in sedimentary rocks.

Many sedimentary rocks contain **fossils** – these are the remains of animals or plants which lived millions of years ago. Most fossils consist of an 'imprint' in a rock: the imprint shows the shape of shells, bones or leaves of the animal or plant which formed the fossil. Such fossils can give us a lot of information about the *age* of the rocks and *where* they were formed.

Certain fossils are very like animals that are alive today. This means we can assume that the animals from which the fossils have come must have lived in a similar way to their modern-day relatives. Even if the fossils of sea shells are found in rocks at the top of mountains, we can be sure that those sea shells must have come from sediments formed from animals which lived at the bottom of the sea.

Going back in time

Other fossils are unlike anything we find today, but they can still give us useful clues. In the south and east of England, the rocks are often quite soft, barely changed from the original sediments. Many of these rocks contain the coiled shells of fossil **ammonites**. To the north and west, however the rocks appear harder and more changed. And here, strange fossil **trilobites** are found. Rocks containing ammonites never lie below those rocks containing tribolites. This means rocks containing ammonites are *younger* than those with trilobite fossils.

Not all the rocks in Britain are of the same age.

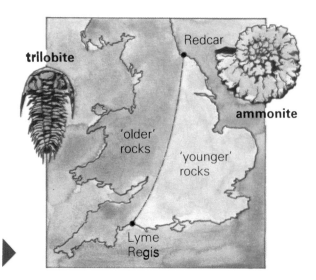

Would you 'date' a fossil?!

By careful study of fossils, the idea of dating the Earth has been extended to produce a standard sequence known as the **stratigraphic column**. Any rock containing fossils can be matched up to this to give the *relative age* of the rock. Life on Earth has changed steadily – **evolved** – over millions of years, so the fossils in each layer act as a 'time fingerprint'. The three main levels are shown here, but these have been further split into hundreds of different levels.

How old is this fossil?

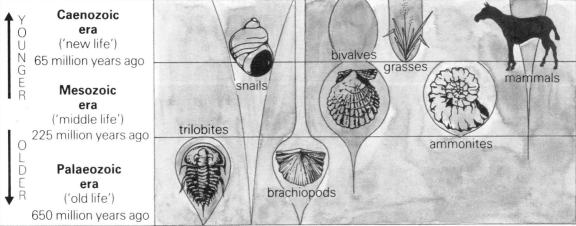

▲ *A stratigraphic column.*

Look closely at fossils – a simple solution to an age-old problem!

A modern way to measure the past

The one thing that fossils cannot do is indicate a precise age in millions of years. But scientists have found a way to do this by studying the **radioactive elements** in the rocks. This study has allowed us to add dates (as shown above) to the stratigraphic column. Radioactive dating also shows that fossils only tell us about part of the Earth's history. The oldest rocks dated in this way have been found to go back to 4000 million years – compare this to the paltry 650 million years age of the oldest rocks with trilobites.

Radioactive dating can give us an age in millions of years, but is very difficult and expensive to do. But a fossil expert could match a fossil to the stratigraphic column in seconds – for nothing! So which is better? It all depends on how accurate you want to be, the period you are seeking to date – and how much money you have got!

1 In a layered sequence of rocks, which is the oldest rock?

2 How is it possible for rocks on hill-tops to contain fossils of sea shells?

3 Give two advantages of radioactive dating.

4 Which Era do these rocks come from?
Rock A contains grass seeds and bones of horse-like mammals.
Rock B contains trilobites and brachiopods.
Rock C contains ammonites, bivalves and snails.

7.11 How time passes!

Any idea of the time?

The Earth has been around for a very long time – 4500 million years! For most of that time, it was very different from the Earth we know today. The idea of millions of years is sometimes difficult to grasp. This scale model of the history of the Earth in a year shows the changes at a rate of just one day for every 12.3 million years!

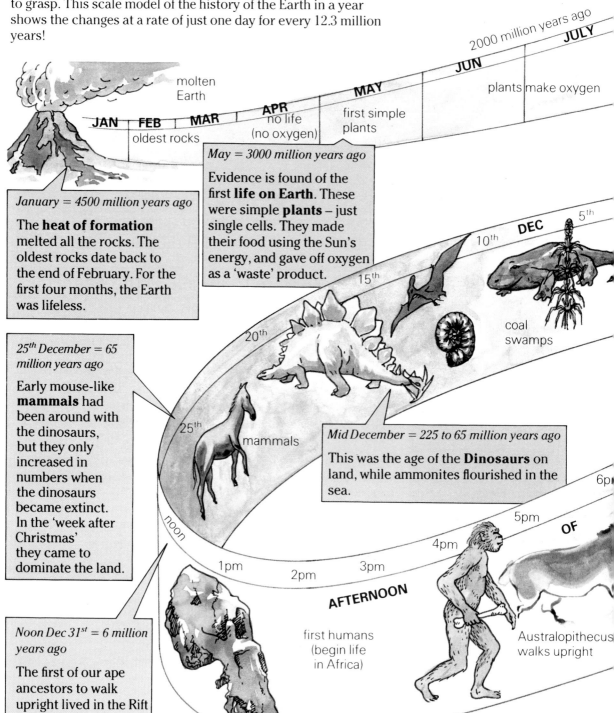

January = 4500 million years ago
The **heat of formation** melted all the rocks. The oldest rocks date back to the end of February. For the first four months, the Earth was lifeless.

May = 3000 million years ago
Evidence is found of the first **life on Earth**. These were simple **plants** – just single cells. They made their food using the Sun's energy, and gave off oxygen as a 'waste' product.

Mid December = 225 to 65 million years ago
This was the age of the **Dinosaurs** on land, while ammonites flourished in the sea.

25th December = 65 million years ago
Early mouse-like **mammals** had been around with the dinosaurs, but they only increased in numbers when the dinosaurs became extinct. In the 'week after Christmas' they came to dominate the land.

Noon Dec 31st = 6 million years ago
The first of our ape ancestors to walk upright lived in the Rift valley of Africa.

October = 1000 million years ago

In this 'safer' world, higher plants developed. **Green algae** appeared. It was like the stuff that grows on the sides of fish tanks. Scientists think that all higher life forms (including us!) come from this.

August = 1500 million years ago

y now **oxygen** had begun o collect in the atmosphere. ome oxygen formed **ozone**, hich stopped dangerous ltra-violet rays from eaching the surface.

November = 600 million years ago

The first animals appeared in the seas – **trilobites** and other strange shellfish. But there was still no life on land.

Late November = 400 million years ago

The first fish had appeared in the seas – strange 'armour-plated' creatures.

ecember = 300 million years ago

ife had now moved onto land. ritain was covered by the **wamp forests** that later ormed our coal.

11.56 pm Dec 31st = 30 thousand years ago

The first humans appeared and quickly spread throughout the world.

midnight = today

Doesn't time fly!

And what of our recorded history? The Romans conquered Britain about 14 seconds before midnight, and 'modern' history (since 1945) can be crammed into the last half-second of the year!

After reading this, you may feel that humans are a little less important in the history of the planet than you once supposed!

7.12 Earth's mineral resources

The Earth's riches

We get all of the materials that our modern technology demands from the Earth's crust. If they were spread out evenly, even the commonest minerals would be difficult to extract. But fortunately, the natural sedimentary processes of **weathering**, **erosion**, **transport** and **deposition** have led to quite pure concentrations of three vital 'bulk' mineral resources: **limestone**, **clay** and **sand**. These resources are quarried (dug out) at certain places in the crust.

Most common elements in crust (as percentage of crust)			
oxygen	50%	potassium	2.5%
silicon	26%	sodium	2.5%
aluminium	7%	magnesium	2%
iron	4%	hydrogen	1%
calcium	3%	carbon	0.3%
		the remaining 83 elements	1.9%

*92 elements occur on Earth, but they are not present in equal amounts. They are usually combined together in various ways to form natural compounds – **minerals**.*

Limestone is used to make cement which is essential for most building work.

Limestone

When clay forms, calcium is released and carried to the sea, where many animals build their shells from calcium carbonate. These shells may be preserved as fossils, or may be broken up to form beds of limestone. This may be quarried and used directly as a **building stone**. Nowadays, limestone is mostly pulverised and roasted to make **cement**. (128 million tonnes is used in this way in Britain every year!)

Clay

Clay forms when igneous rock weathers. It may stay in place to form **soil**, or be washed away to form beds of clay. 6 million tonnes of clay are still dug up every year in Britain, nearly half of which is used for **brickmaking**. Purer clays are used for **pottery** and **china**.

Often unsightly disused clay pits can be 'reclaimed' to help restore the environment again.

Sand

Grains of **silica** (silicon dioxide) are released when granite weathers. The silica may be washed away to collect as beds of **sand** in rivers or the sea. Sand and gravel are used in vast quantities as **aggregates** – fillers for concrete and other construction materials. Every year, 160 million tonnes of sand are used in Britain.

As quartz is so resistant, it may be reworked by sedimentary processes many times, each time becoming purer and purer. The purest sand is used to make **glass**.

The glass used to makes these bottles needs very pure sand (or recycled glass!).

What about coal?

Coal is another important resource – almost pure carbon from the remains of plants that covered Britain 300 million years ago. You can find out more about coal on 4.11).

... and the sea's salts?

After the coal formed in Britain, the climate changed, becoming hot and desert-like. The shallow sea that had covered this area started to dry up. Salt deposits formed – these were called **evaporites**. One example of an evaporite site is the Great Salt Lake of Utah, USA.

Evaporites are useful sources of chemicals such as sodium chloride, magnesium chloride and magnesium sulphates.

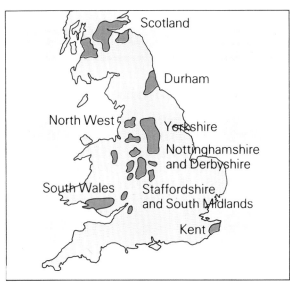

These are the main coalmining areas of Britain. We have a lot of coal, but even so, it will only last for 300 years.

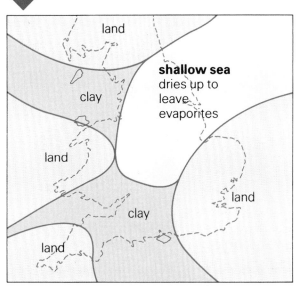

250 million years ago, Britain did not exist as we now know it. Instead there was a shallow sea that dried up, leaving thick deposits of gypsum and rock salt.

Rock salt (sodium chloride): Three million tonnes of rock salt are mined every year. As well as for sprinkling on chips, salt is used as a raw material for making sodium and chlorine. But it is mostly used straight from the mine – rock, clay and all – on roads in winter to **melt** ice.

Potassium and **magnesium salts** are rare in Britain, so most are imported for use in the **chemical industries**, and potassium salts as **fertilisers.**

Gypsum (calcium sulphate): Four million tonnes of gypsum are produced every year in Britain. This is crushed and heated to make plaster of Paris – mostly for the building trade. It also has an important use as blackboard 'chalk'!

What about the metal ores?

Metals are so important, that they have a section of their own. You can find out about the metal ores on 6.8 and 6.11).

1 Draw a bar chart to show the amounts of the major 'bulk' minerals produced in Britain every year.

2 Over the last 30 years, the proportion of sand and limestone used has gone up compared to clay. What change in building techniques might have caused this?

3 How have animals and plants helped to concentrate mineral resources?

4 If an inland sea 100 m deep evaporated, you would only get 1 metre of salt, yet many salt beds are much thicker. How could this happen without having a deeper sea in the first place?

7.13 How to identify minerals

It's not that difficult

Though there are hundreds of different minerals to be found in the crust of the Earth, most rocks are built from just a handful of common 'rock-forming' minerals. Similarly, in mineral veins all over the Earth, the same metal ores crop up over and over again. This means only a few tests are needed to identify most ores. The common minerals can usually be identified using a few simple tests . . .

Simple observation is often enough to identify minerals.

Lustre

This is how the mineral looks. Most common minerals either seem glassy or metallic.

Hardness

This is a very useful test. Scientists use a **scale of hardness** from *one* (talc – the softest) to *ten* (diamond – the hardest). We can make do with just four tests – whether the mineral can *scratch glass*, or whether it can itself be *scratched by a steel file, a copper coin*, or *a fingernail*. ▼

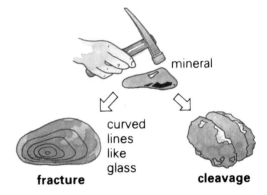

A simple breakage test.

Breakage

Some minerals break in a very irregular way, others break along curvy lines, like glass. Many minerals break along flat, mirror-like surfaces called **cleavage planes.**

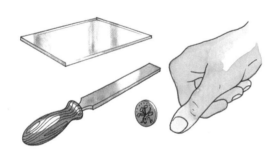

All you need for a hardness test.

Streak

The colour of a mineral can be misleading, but the colour when *powdered* is much more useful. This is best seen by scratching the mineral across the back of a special tile – leaving a '**streak**' of powder.

Density

On average, the rocks of the Earth's crust have a density about *two and a half* times that of water, and most minerals are about the same. But some ore minerals are two or three times as dense as this, and feel distinctly heavy for their size.

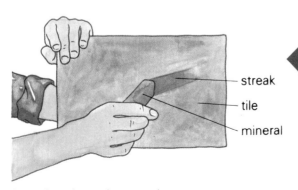

Check for colour using a streak test.

Mineral	Hardness	Lustre	Breakage	Streak	Density	Other information
Quartz	cannot be scratched by a file	glassy	curvy	white	average	A common rock-forming mineral. Found in granite and sandstone. Used to make glass.
Felspar	scratches glass – but scratched by a file	glassy	cleavage	white or pinkish	average	A common rock-forming mineral. Often forms large pink crystals in granite.
Calcite	scratched by a coin – but not by a fingernail	glassy	cleavage	white	average	Common in sedimentary rocks. Forms stalactites in caves. Fizzes with acid.
Galena	can be scratched by fingernail	silver-metallic	cleavage	grey to black	very high	Lead ore (lead sulphate). Found in veins.
Pyrites	scratches glass – but scratched by a file	brassy metallic	variable	green-black	high	Fools' gold – it sometimes fooled the old miners! Iron ore (iron sulphide).

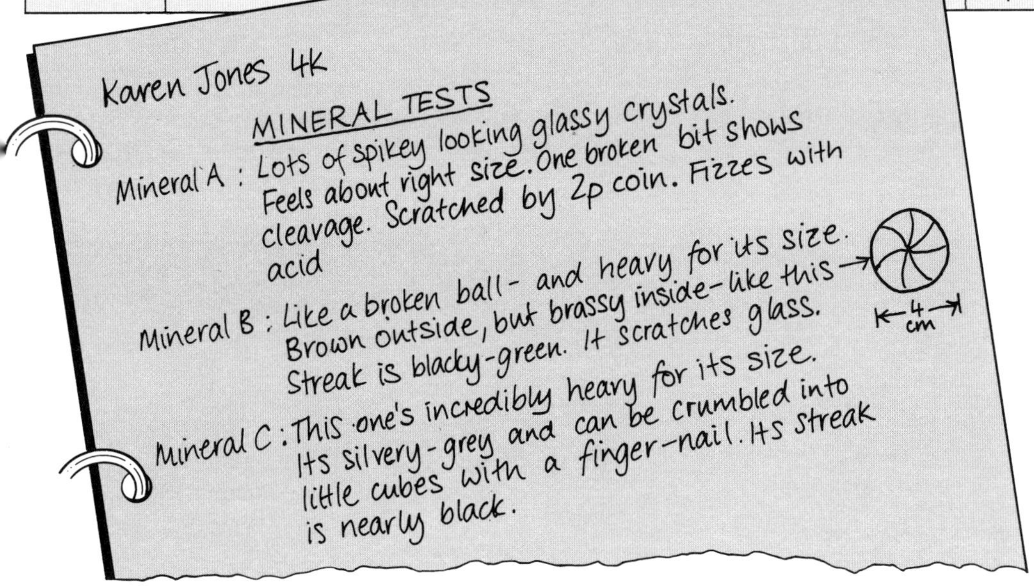

Karen Jones 4k

MINERAL TESTS

Mineral A : Lots of spikey looking glassy crystals. Feels about right size. One broken bit shows cleavage. Scratched by 2p coin. Fizzes with acid

Mineral B : Like a broken ball – and heavy for its size. Brown outside, but brassy inside – like this → Streak is blacky-green. It scratches glass. ← 4 cm →

Mineral C : This one's incredibly heavy for its size. Its silvery-grey and can be crumbled into little cubes with a finger-nail. Its streak is nearly black.

1 Karen used these tests to describe some mineral specimens. She wrote her results in her notebook as shown above. Use the table of mineral properties to identify minerals **A** to **C**.

2 Crystals of quartz and calcite can look very similar. Describe three differences in their properties that could be used to tell them apart.

3 Iron pyrites is usually brassy-yellow but one form (marcasite) is more silvery. How could it be tested to show that it was not galena?

7.14 Armero: an avoidable tragedy?

Why live in Armero?

Armero stood in what had been a tranquil, tree-lined valley. The rich volcanic soil provided an abundance of food and the area had escaped disaster for generations. Like their counterparts in Pompeii nearly 2000 years ago, the local people did not take seriously the threat from the nearby sleeping giant... the volcano Nevada del Ruiz.

At 5000m, the volcano was snow-capped, but this snow melted rapidly in the heat of the eruption, pouring water into the steep river valley.

Much of this volcano was built from loose **volcanic ash** and was as unstable as a pile of dry sand. The flooding river easily scoured out this material.

The Andes mountains were formed as the Pacific Ocean plate pushed into and under the South American plate. This collision has forced up new **fold mountains**, and created enough heat to melt great masses of the lower crust. Some of this **molten magma** has forced its way to the surface, forming explosive **volcanoes** such as Nevada del Ruiz.

Minor **earthquakes** associated with the eruption made part of the valley side collapse. This **temporarily dammed** part of the river, allowing the floodwaters to build up further.

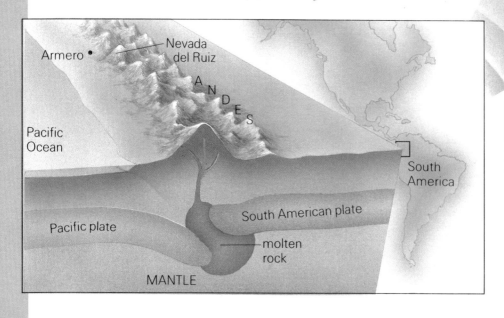

Columbian Catastrophe
20 000 DEAD

On a November night in 1985, 25 000 people settled down for the evening in Armero, a town at the northern end of the Andes mountains of South America. They knew that Nevada del Ruiz – the volcano that stood 50 km away – was active. Ash had fallen a couple of months before and, more recently, lava had been reported flowing from the crater. But they had listened to their local radio and government officials had assured them that they were perfectly safe. They were not too concerned, even when they heard the volcano erupt . .

. . . Shortly afterwards, a torrent of mud drove through Armero at 100 km per hour, killing 20 000 people. By morning, all that could be seen was the corrugated iron of the roofs sticking up through the mud and slime.

The swollen river soon **burst** through, carrying millions of tons of mud and ash in its turbulent flow.

As it hit the flatter ground near Armero, it spread out, losing its power and dropping its load of mud – right on Armero.

A disaster just waiting to happen?

Before the fateful night, there had been plenty of warning of the trouble to come. The tremors and minor eruptions had been monitored for months. The danger from loose ash was well known – such mudslides are common around volcanoes in the Andes. Yet no-one in authority had put these two pieces of information together, so no-one realised the potential for disaster. This meant there was no move to evacuate the area.

Who, if anyone, was to blame? And what would you have done if you were there and had seen the warning signs? You may find it helpful to look back through this module before you answer these questions.

ARMERO

MODULE 7 EARTH SCIENCE

Index
(refers to spread numbers)

A
acid rain 7.8
active volcano 7.3
African plate 7.5
aggregates 7.12
Alps 7.5
ammonites 7.5
Antarctic plate 7.5
Armero 7.14
ash (volcanic) 7.3

B
bedding planes 7.9
binding action 7.8
breakage 7.13
brickmaking 7.12

C
Caenzoic era 7.10
calcite 7.13
calcium carbonate 7.8
calcium sulphate 7.12
carbon dioxide 7.8
chemical attack 7.8
china 7.12
clay 7.12
cleavage planes 7.11
conglomerate 7.6, 7.9
core 7.2
crater 7.3
crust 7.2

D
deposition 7.9

E
earthquakes 7.1, 7.2, 7.4
epicentre 7.2
erosion 7.7.7, 7.9, 7.12
eruption 7.3, 7.14
evaporites 7.12
evolving 7.10
Eurasian plate 7.5

F
faulted 7.10
felspar 7.13
fertilisers 7.12
flood plains 7.9
folded 7.10
fold mountains 7.5

G
glaciers 7.7
gneiss 7.6
gorge 7.7
granite 7.6
gypsum 7.12

I
igneous rocks 7.6

L
lustre 7.13

M
mammals 7.11
mantle 7.2
Mesozoic era 7.10
metamorphic rocks 7.6, 7.
metamorphosis 7.8
Mexico city 7.1, 7.4
Mid atlantic ridge 7.5
molten rock 7.5
mountains 7.5

N
North American plate 7.5

P
pacific plate 7.5
Palaeozoic era 7.10
plate tectonics 7.5
plates (continental) 7.5
pottery 7.12
pyrites 7.13

Q
quartz 7.13

R
radioactive elements 7.10
rain forests 7.8
Richter scale 7.4
Ring of Fire 7.5
rivers 7.9
rock cycle 7.6
rocks 7.6, 7.7

S
salts
   magnesium 7.12
   potassium 7.12
San Andreas fault 7.4
sand 7.12
sand stone 7.9
sedimentary rocks 7.6
seismometer 7.2
shells 7.8
silica 7.12
soil 7.8
South American plate 7.5
stalactites 7.8
stalagmites 7.8
stratigraphic column 7.10
streak test 7.13
swallow holes 7.8
swamp forests 7.11

T
transport 7.7
trilobite 7.10

V
valleys 7.7

W
weathering 7.12

Photo Acknowledgements

The references are indicated by spread numbers and, where appropriate, by photo sequence.

Ancient Antiques Architecture Collection (Ronald Sheridan) *7.3/2*; Associated Press *7.1/1, 7.1/3*; Bruce Coleman (Alain Compost) *7.8/1*; Sally & Richard Greenhill *7.12/4*; GeoScience Features *7.2, 7.5/2, 7./6/2, .6/3, .6/4*; Colin Johnson *7.5/1*; Frank Lane Agency *7.1/2* (Thomas Micek) *7.3/1*, (S Mccutcheon) *7.4/1*, (M Nimmo) *7.7/4, 7.9/1, 7.9/2*; Science Photo Library (NASA) *7.9/3*; South American Pictures (Nicholas Bright) *7.14/2, 7.14/3*; Martin Stirrup *7.7/5, 7.7/7, 7.9/4, 7.9/5, 7.9/7*; Topham Picture Library – contents, *7.8/2, 7.12/1, 7.12/2, 7.7/3, 7.7/6, 7.9/6, 7.10/1*; United Glass *7.12/3*; Waltham *7.4/3, 7.7/1, 7.10/2*

Picture research: Jennifer Johnson